Urban Tree Management

MW00844848

Urban Tree Management

for the Sustainable Development of Green Cities

EDITED BY

Andreas Roloff
Dresden University of Technology,
Tharandt, Germany

with contributions by:
Eckhard Auch, Markus Biernath, Sten Gillner, Mathias Hofmann, Doris Krabel,
Rolf Kehr, Sandra Korn, Matthias Meyer, Ulrich Pietzarka, Hubertus Pohris,
Jürgen Pretzsch, Andreas Roloff, Steffen Rust, Andreas Tharang, Juliane Vogt

Dresden University of Technology, Germany

WILEY Blackwell

This edition first published 2016 © 2016 by John Wiley & Sons, Ltd.

Registered Office
John Wiley & Sons, Ltd., The Atrium, Southern Gate, Chichester, West Sussex, PO19 8SQ, UK

Editorial Offices
9600 Garsington Road, Oxford, OX4 2DQ, UK
The Atrium, Southern Gate, Chichester, West Sussex, PO19 8SQ, UK
111 River Street, Hoboken, NJ 07030-5774, USA

For details of our global editorial offices, for customer services and for information about how to apply for permission to reuse the copyright material in this book please see our website at www.wiley.com/wiley-blackwell.

The right of the author to be identified as the author of this work has been asserted in accordance with the UK Copyright, Designs and Patents Act 1988.

All rights reserved. No part of this publication may be reproduced, stored in a retrieval system, or transmitted, in any form or by any means, electronic, mechanical, photocopying, recording or otherwise, except as permitted by the UK Copyright, Designs and Patents Act 1988, without the prior permission of the publisher.

Designations used by companies to distinguish their products are often claimed as trademarks. All brand names and product names used in this book are trade names, service marks, trademarks or registered trademarks of their respective owners. The publisher is not associated with any product or vendor mentioned in this book.

Limit of Liability/Disclaimer of Warranty: While the publisher and author(s) have used their best efforts in preparing this book, they make no representations or warranties with respect to the accuracy or completeness of the contents of this book and specifically disclaim any implied warranties of merchantability or fitness for a particular purpose. It is sold on the understanding that the publisher is not engaged in rendering professional services and neither the publisher nor the author shall be liable for damages arising herefrom. If professional advice or other expert assistance is required, the services of a competent professional should be sought.

Library of Congress Cataloging-in-Publication Data

Urban tree management : for the sustainable development of green cities / edited by Andreas Roloff.
     pages   cm
   Includes index.
   ISBN 978-1-118-95458-4 (pbk.)
1. Urban forestry. 2. Trees in cities. 3. Sustainable development. I. Roloff, Andreas, 1955– editor.
   SB436.U736 2016
   634.9–dc23
                                        2015031844

A catalogue record for this book is available from the British Library.

Wiley also publishes its books in a variety of electronic formats. Some content that appears in print may not be available in electronic books.

Cover image: © Andreas Roloff

Set in 9.5/11.5pt Meridien by SPi Global, Pondicherry, India

Printed in Singapore by C.O.S. Printers Pte Ltd

1   2016

Contents

List of contributors

Dr. Eckhard Auch

Technische Universität Dresden, Institute of International Forestry and Forest Products, Tharandt, Germany

Dr. Markus Biernath

Staatsbetrieb Sachsenforst, Forest District Dresden, Dresden, Germany

Dr. Sten Gillner

Technische Universität Dresden, Institute of Forest Botany and Forest Zoology, Tharandt, Germany

Dr. Mathias Hofmann

Swiss Federal Institute for Forest, Snow and Landscape Research WSL, Social Sciences in Landscape Research Subunit, Birmensdorf, Switzerland

Prof. Dr. Rolf Kehr

HAWK Hochschule für Angewandte Wissenschaft und Kunst, Fakultät Ressourcenmanagement, Göttingen, Germany

Dr. Britt Kniesel

Technische Universität Dresden, Institute of Forest Botany and Forest Zoology, Tharandt, Germany

Dr. Sandra Korn

Technische Universität Dresden, Institute of Forest Botany and Forest Zoology, Tharandt, Germany

Prof. Dr. Doris Krabel

Technische Universität Dresden, Institute of Forest Botany and Forest Zoology, Tharandt, Germany

Dr. Matthias Meyer

Technische Universität Dresden, Institute of Forest Botany and Forest Zoology, Tharandt, Germany

Dr. Ulrich Pietzarka

Technische Universität Dresden, Tharandt Botanic Garden & Arboretum, Tharandt, Germany

Dr. Hubertus Pohris

Technische Universität Dresden, Institute of International Forestry and Forest Products, Tharandt, Germany

Prof. Dr. Jürgen Pretzsch

Technische Universität Dresden, Institute of International Forestry and Forest Products, Tharandt, Germany

Prof. Dr. Andreas Roloff

Technische Universität Dresden, Institute of Forest Botany and Forest Zoology, Tharandt, Germany

Prof. Dr. Steffen Rust

HAWK Hochschule für angewandte Wissenschaft und Kunst, Fakultät Ressourcenmanagement, Göttingen, Germany

Andreas Tharang

Technische Universität Dresden, Institute of Geodesy, Dresden, Germany

Dr. Juliane Vogt

Technische Universität Dresden, Institute of Forest Biometry and Systems Analysis, Tharandt, Germany

CHAPTER 1

Intro: Urban trees – Importance, benefits, problems

Andreas Roloff

Technische Universität Dresden, Tharandt, Germany

1.1 Introduction

Trees often and quickly gain a bad reputation, caused by falling branches or entire trees, roots in sewage drains, neighbors fighting over fruit and leaves littering their gardens, health issues from pollen allergies, etc. The problems caused by city trees are usually more conspicuous and have greater ramifications. Their advantages can often be difficult to record and to assess. As a result, the negative impacts are much more widely discussed, whereas extensive papers about their positive aspects are rare.

This chapter, therefore, aims to raise awareness of the positive impacts and benefits of urban trees and their importance to city dwellers (Figures 1.1 and 1.2). It describes their advantages (with no claim to completeness) and details their effects on our quality of life and well-being – aspects that are increasingly important in these times of progressing urbanization.

"If I knew the world would end tomorrow, I would still plant another tree today."

– Martin Luther

1.2 Aesthetics, sensory impressions

To many people, the *beauty of nature* is manifest in trees (Tyrväinen *et al.*, 2005). This is especially true for ancient, free-standing trees. Their variation in phenology (change of appearance across the seasons), including shooting, blooming, fruit, leaf coloring and falling leaves, is an important factor in how we *experience the seasons*, especially in cities. Many trees even change their smell over the course of the year. Areas without trees can be areas without seasons, especially in temperate climates.

Visual impressions such as coloring (e.g., of the leaves in spring and autumn), different structures (e.g., the shape of the leaves and the architecture of the treetops), design (e.g., Lombardy poplars, ancient oaks) and aesthetics (how a tree affects us) cause positive emotions and experiences (Bahamón, 2008; Miller, 2007; Trowbridge and Bassuk, 2004; Smardon, 1988; Velarde *et al.*, 2007). As an example of the *aesthetic impact* of different tree species, just think of a light, young grove of birches, as opposed to a dark, dense forest of conifers in spring. An assessment based on aesthetics is, of course, subjective, but it may still be used, for example, to rank city trees by popularity.

Urban Tree Management: For the Sustainable Development of Green Cities, First Edition. Edited by Andreas Roloff.
© 2016 John Wiley & Sons, Ltd. Published 2016 by John Wiley & Sons, Ltd.

Figure 1.1 Treeless square – cold, hard, unwelcoming, and easily overheated in summer.

Figure 1.2 Square with trees (*Fraxinus angustifolia*) – segmented, warm, inviting and shady.

   Aside from visual impressions, the senses of smell (blossoms, autumn leaves), hearing (rustling of the treetops, rustling of the fallen leaves), taste (fruit, young leaves) and touch (fruit, young leaves) also play important roles.

1.3 Psychology, well-being, health

Trees accompany us through life. Relationships between trees and people are complex and have been poorly investigated. The potential of such relationships becomes clear if you consider the *"house tree"*. Even today, it is not uncommon for families to plant a tree next to their home – for example, to serve as a "patron", in order to have shade in summer and shelter from wind, or in order to grow fruit, honey, and so on (Figure 1.3). Some house trees are even considered a member of the family, and the bond is particularly strong if the tree was planted by the owner of the home to mark a special occasion. Positive feelings towards house trees are usually associated with aesthetics: "looks nice"; "the blossoms"; "the color of the leaves". Gardens often contain many of the "public" trees along the streets and in the parks of a city. In Dresden, for example, there are 600,000 private garden trees, but only 60,000 public street trees (Roloff, 2013).

   Trees are also increasingly important for our health – for instance, visits to parks (municipal, spas and civic parks) and gardens, walks and hikes, resting on a bench under a tree, picnics in

Figure 1.3 House tree (*Acer pseudoplatanus*) – often a member of the family.

Figure 1.4 Relaxing under a tree (*Castanea sativa*) in a park in Cornwall, UK.

the shade of trees (a popular custom in many cultures). Parks may therefore also be called "therapeutic landscapes", and a movement called "garden therapy" is currently on the rise. Gardens (including allotments) are increasingly seen as personal spas, as a place where people can feel comfortable and relax – gardening as private health care.

In addition, city trees also protect us from *emissions*, especially by reducing the levels of ozone, nitrogen oxides, sulfur and carbon dioxide (Harris *et al.*, 2004; Konijnendijk *et al.*, 2005; Tyrväinen *et al.*, 2005; Yang *et al.*, 2005; Donovan *et al.*, 2011). Parks act as a city's "green lungs". In recent years there have been many discussions about *particulate matter* and how it can be reduced to protect our health, with a focus on the ability of trees to bind microparticles in their leaves. Benefits from this depend on factors such as the placement of the trees along the streets, and the width of the streets (see Chapter 13).

Due to their positive impact on our psyche and health (Harris *et al.*, 2004; Konijnendijk *et al.*, 2005; Tyrväinen *et al.*, 2005; Arnberger, 2006; Hansmann *et al.*, 2007; Carreiro *et al.*, 2008; Konijnendijk, 2008; Miller, 2007; Velarde *et al.*, 2007; Cox, 2011; Lee and Maheswaran, 2011), and because they have been proven to accelerate recovery and regeneration, trees often dominate the parks that belong to spas, asylums, or hospitals, as well as cemeteries. They also provide the shade needed in summer, they reduce noise and improve the quality of the air, and they have a calming effect on the mind (see Figure 1.4) (Harris *et al.*, 2004; Tyrväinen *et al.*, 2005). Parks are also popular places for physical activities (ball games, walking, running, etc. – Lohr *et al.*, 2004; Matsuoka and Kaplan, 2008). Recent research shows the importance of nature for the living environment and local recreation; nature is

Table 1.1 Psychological, physical and social benefits of being and exercising in urban parks.

Psychological benefits	• being close to nature brings us closer to our own nature • increased relaxation and well-being • reduced stress levels • anti-depressant effects of light and greenery
Physical benefits	• from physical activities (walking, exercising, etc) • increased energy from cool and fresh air • stimulation of all the senses • increased cognitive performance
Social benefits	• stimulation of interaction, especially between children and adults, or between different cultures • increased feeling of belonging, reduced social isolation • increased feeling of responsibility due to consideration of natural interrelations.

seen as the most important factor. The physical, psychological and social benefits of being and exercising in green areas (e.g., in parks and gardens) are summarized in Table 1.1.

Trees can also have a lasting influence in childhood, for instance by forming a local identity; a tree species that dominated our surroundings in childhood is usually associated with fond memories, and we may love it for life. *Tree adoptions* are a popular gift, and usually result in a personal relationship between the presentee and the tree or the tree species. Planting a *birth tree* at the birth of a child used to be common practice (and is becoming popular again). In the 18th century, many places had a law that a wedding license would only be granted if a certain number of young, verdant *wedding trees* had been planted. Wedding avenues and dummy trees (to help children wean off their pacifiers – see Figure 1.5) are examples of modern customs associated with trees.

In past times, *dance and court trees* used to be very important. Dance lindens had a platform in the crown where dances were held; court lindens were used for meetings dedicated to law and order. The finding of justice was based on the belief that nobody would dare lie under a *Tilia* tree.

Maypoles are tall, pruned and decorated trees that are raised as part of a festive celebration in a central place in the town. A *roofing ceremony* is a celebration under a tree that is attached to the roof when the shell of the building has been completed (Figure 1.6).

Trees are also central in *landscaping*, such as in the planning and construction of parks, squares, private and landscaped gardens. A park that resembles a savannah (Figure 1.7) is particularly beneficial for humans. It accommodates our primal urge for keeping everything in sight, which originates from the prehistoric development of humankind in the African savannah. Looking at trees and shrubs gives us pleasure. Trees can create a certain ambiance (e.g., potted palm trees for a tropical flair).

The psychological aspects of the relationship between people and trees are particularly noticeable in tree-based *horoscopes* (e.g., the "Celtic tree horoscope"). These use certain tree types, depending on their appearance (e.g., *Quercus* for toughness, *Pinus* for pickiness, *Salix* for melancholia).

Because ancient, giant specimens have always awed humans, trees also play an important role in mythology. Trees are the most suitable image to represent humanity, because they, too, stand tall and raise their "arms" toward heaven ("Trees are like brothers"). Many religious scriptures have tree allegories, and many sayings also use the simile of tree and human, e.g., "a bad tree does not yield good apples" or "the apple doesn't fall far from the tree". Many places have sacred woods.

Figure 1.5 Dummy tree (*Liriodendron tulipifera*), to help children wean off their pacifiers.

Figure 1.6 *Betula pendula* tree in a roofing ceremony – a decoration very popular in house building.

Figure 1.7 A park that has been landscaped to resemble a savannah, including stretches of lawn with individual trees and edges of woods.

Aspects of ancient religions often revolve around trees, such as the "tree of forbidden knowledge" at the beginning of the Bible, when Adam and Eve are expelled from Paradise after eating its fruit. Other examples are the giant ash Yggdrasil in Norse religion, or Buddha's Tree of Enlightenment. Even today, the symbolic and spiritual importance of trees can be deduced from the coins of many countries that show trees or leaves. *Songs, literature, poetry and fairy tales* often revolve around trees. Many ancient stories tell of people who are turned into trees.

Last, but not least, there are trees that are very special. In Germany there is one that is very special for relationships – an old *Quercus* tree, close to the village Eutin (north of Hamburg), called the "Flirt tree". It is the only tree in Germany that has its own postal address and its own "letterbox" (a hole in the trunk; Figure 1.8) – exempt from the sanctity of mail. People write to the tree about their wish for a partner, husband or wife, or read and respond to the letters left in the tree by others.

Aging in trees is usually seen as something positive; the older a tree, the bigger the impression it makes. Ancient trees represent birth and decline, give us a feeling of timelessness and connect us with past eras (Luther's *Tilia*, Goethe's *Gingko*, Newton's apple tree – Stokes and Rodger, 2004). At the same time, they make us aware of the modest role and lifespan of the individual person. Trees create an atmosphere of piece and quiet, thus helping us to *relax and improving our moods*. City dwellers in Michigan, USA, said in a survey that trees are the strongest contributing factor for the attractiveness of streets and districts (Figure 1.9), whereas their absence was the most negative factor: "Streets without trees have no face."

Figure 1.8 The "letterbox" of the "Flirt tree" (*Quercus robur*), used for searching for a partner.

Figure 1.9 Road-side trees (*Acer platanoides*) with a strong positive effect on the quality of living.

Color psychology studies the effect of colors on the psyche. According to its position in the chromatic circle, the green of leaves has a balancing and calming effect. It induces harmony, inspires, stabilizes, improves our self-esteem and makes us yearn for (the lost) paradise. This is another reason why woods and parks have such a relaxing effect on us.

Green is particularly beneficial for people who are prone to mood swings. Going for a walk in the forest or the park usually has great a very positive effect on people with depression. Green is also the color traditionally used by environmentalists, because it represents life and a healthy environment.

The so-called *"Tree test"* – the interpretation of spontaneously drawn tree pictures – has been a much-used (albeit controversial) method in psychotherapy for finding causes for abnormal behavior in children and adolescents. Therapists analyze the centre of gravity and the crowning height of the tree, as well as peculiarities in the rendition of trunk, crown and roots.

1.4 Environmental education, ecology

Trees are essential for encouraging *environmental awareness* in cities. City dwellers are increasingly alienated from nature, but trees allow them to experience a small measure of wilderness and of the wonders of nature by realizing the principles of adaptation, optimization and sustainability (Kowarik and Körner, 2005; Miller, 2007; Konijnendijk, 2008; Hofmann *et al.*, 2012), and by changing and reacting in the course of the year (and over the years). They also play an important role in the environmental education of children. Proof of this are the many recently established forest kindergartens.

Trees contribute to *biodiversity, conservation of nature and preservation of genes* (Tommasi *et al.*, 2004). Moreover they form the habitat for birds (e.g., jay), insects (e.g., longicorn), mammals (e.g., squirrel), epiphytes (e.g., mistletoe), fungi (e.g., tinder fungus), lichens (e.g., common orange lichen) and so on, and help to integrate and link biotopes across parks, green corridors or avenues.

1.5 Orientation, spacious ordering, architecture

Avenues and tree-lined streets and roads have been used for centuries (in some cases, even for millennia) for *orientation and guidance*. They direct the eyes or show the way (e.g., toward important buildings or prominent locations), increasing road safety at the same time.

Trees also contribute to the *enhancement, structuring and design* of public open spaces by separating them into individual, yet not entirely disconnected areas, increasing the impression of space (Figure 1.2). Ancient trees in squares and significant places in towns and cities are often a decisive factor in *forming the townscape*, as can be seen in place names such as "Royal Oak", "Elm Tree" (also common pub names) or "Hollywood". Such markers are often protected as natural monuments; in some cultures, they are even considered sacred (e.g., in China).

In (landscape) architecture, trees have long been used for purposes such as directing the view, for emphasizing the shape and style of buildings, as a framework, as a contrast, for creating a connection to the gardens or the surrounding landscape (e.g., in Singapore), and also in their role as house trees.

1.6 Protection, quality of life

Trees are currently gaining great importance in *local climate protection*, as rising temperatures make their role in providing shade and increasing humidity (by transpiring) more and more relevant (Harris *et al.*, 2004; Heidt and Neef, 2008; Konijnendiek, 2008;

Figure 1.10 Desired shading by urban trees (*Liquidambar styraciflua*) on a hot summer day.

Bowler *et al.*, 2010). City trees equalize extreme temperatures. They contribute to cooling and shading (Figure 1.10), which is perceived as pleasant. While temperature differences between parks and areas covered by buildings or concrete may be measured at up to 5°C, the perceived difference in physiologically equivalent temperature (PET) is usually much higher and may reach more than 10°C (due to the increased humidity under trees). The difference in surface temperature between asphalt and tree-covered greens is even more extreme (up to 15°C).

The density and surface area of the leaves is, of course, an important determining factor in this, and individual trees are far less effective than groves and woods. Every unit of LAI (Leaf Area Index, measuring the density of the foliage) reduces the surface temperature of the shaded area by approx. 1°C on hot summer days (Hardin and Jensen, 2007).

Current discussion about global warming assigns additional importance to trees because they are able to *fix carbon dioxide*. This may be relevant not only on a local scale for calculating the carbon footprint of a plot of land or of an entire city, but also on a global scale for scenarios of further global warming.

In *noise protection*, trees are important not only because of the objectively measurable reduction of noise by up to 10 dB, but also because of their psychological effect; they block traffic from view, which makes the reduction of the noise feel greater than it actually is (Bucur, 2006; Miller, 2007). How efficiently a treetop works as a *screen* depends on the tree's age and on the LAI (Leaf Area Index) of its species. Dense branching with a great number of small leaves is particularly effective (see Chapter 13).

Trees also provide good *shelter from wind* (Coutts and Grace, 1995; Harris *et al.*, 2004; Trowbridge and Bassuk, 2004). In windy regions, it is therefore common to plant and cultivate rows of trees (e.g., along cycle paths). For free-standing houses, protection from wind is one of the most important capacities of house trees; in damp or wet isolated locations, they are also important as *protection from lighting*.

In *bioengineering*, trees (especially willows and alders) are important as slope protection and for erosion control. In densely populated areas, trees contribute significantly to *water pollution control*, rain water retention, and flood control (e.g., in spring reserves).

1.7 Food/diet, healing powers

Pomaceous fruit (e.g., apples, pears, cherries, peaches) and nuts (e.g., hazelnuts, walnuts) are an *integral part of our diet*. Their advantage is that, with regards to pathogens, trees are at a far lower risk than agricultural crops. In addition, they usually do not deteriorate the soil, and so do not require fertilizing. Mushrooms should also be mentioned in this context – many species are mycorrhizal fungi, and therefore depend on trees.

Some species are important for bees, and are thus involved in the production of honey. Leaves can be used to make tea (e.g., camellia). Connoisseurs also use them in salads. Leaves were also traditionally used (and still are) as fodder.

Many tree substances have *medical benefits* that still play a very important role in many cultures. Phytotherapy (botanical medicine) uses both traditional tree supplements (e.g., gingko substances for improving blood circulation) and new discoveries (e.g., cancer treatment based on yew substances) (Clarke, 1996).

1.8 Utilization of trees

Children like building *tree houses* and use horse chestnuts and acorns for making figurines. *Playgrounds* and gardens often boast trees for climbing and swings.

Wood is used to make furniture and utensils or is used just as firewood. Leaves and bark can be used to dye natural materials. The spring sap of maples (especially sugar maple) can be made into syrup (maple syrup) and is an important ingredient for pancakes, ice cream and other dishes; the spring sap of birches can be used for hair tonics; the bark of cork oaks is used as an insulating material and for corking bottles.

Mention must also be made here of the *Christmas tree*, which has been a popular custom in many countries for 500 years.

1.9 Economic and social advantages

Trees are an *economic factor*, because people prefer green cities and districts. However their advantages are difficult to calculate from an economical and monetary perspective (Schulz and Balder, 2000; Price, 2003; Harris *et al.*, 2004; Konijnendijk *et al.*, 2005; Miller, 2007; Carreiro *et al.*, 2008; Konijnendijk, 2008). In order to get a rough idea of the value of the advantages of city trees, compare them to appropriate substitutes:
• fruit crops vs. supermarket fruit;
• shading from trees vs. parasols or blinds;
• trees screening people from view vs. fences or walls;

• the cooling effect of foliage vs. air conditioning (Figure 1.10);
• air purification by trees vs. technological filters.

The economic value of trees for the United States, for example, has been estimated at 3.1 billion Euros per year, based on their emission control alone. The relationship between costs and benefits of city trees was calculated for two Californian cities, with cost/benefit ratios of 1.8 and 1.5 – an interesting result that clearly shows that the advantages prevail (McPherson and Simpson, 2002; Nowak *et al.*, 2006). Real estate agents often use proximity and the availability of green areas for advertising. Trees, especially older trees, can even *increase property value*. Mansions are incomplete without an older tree population. City trees also have an indirect economic value, such as in beer gardens or open-air restaurants, or for tourism.

In East Asia, it has been popular for a long time to exercise or meditate together in parks, as part of the morning relaxation ritual. The concept of the *"green gym"* is spreading rapidly all over the world. The same development is happening with guerrilla gardening: the planting of plants and trees in city centers by private people at their own cost. City dwellers want to experience more of nature in the city, together with like-minded people. At the same time, they can do something beneficial for themselves and for *urban greening*. This attracts young and old, rich and poor, and brings them together.

Parks in problem districts are especially suited to this type of physical activity, as it helps reduce prejudices against "the others", and participants from the area become increasingly conscious of the value of "their" green area. Intercultural exchanges and acceptance also benefit. Public woods and parks therefore provide a socio-cultural dimension by combining cultural processes and social networks (Harris *et al.*, 2004; Konijnendijk *et al.*, 2005; Tyrväinen *et al.*, 2005; Sanesi *et al.*, 2006; Heidt and Neef, 2008; Konijnendijk, 2008; Secco and Zulian, 2008; Sugiyama and Thompson, 2008). The provision of places for social interaction brings *social advantages*: meetings and events such as picnics with friends, a concert in the park or a cherry blossom festival; chance encounters on park benches; or when taking a pram for a stroll or a dog for a walk.

1.10 Issues

Finally, we would like also to mention the well-known and much-discussed negative issues that may be caused by urban trees, for example because of:
• fruits, leaves;
• vine lice, resin;
• pollen (allergies);
• falling branches;
• falling trees;
• undesired shading in summer;
• damage to pipelines from roots;
• damage to buildings from roots;
• bird droppings;
• raised cobblestones;
• accidents with trees (collisions);
• restrictions to building construction due to tree protection legislation;
• costs for tree maintenance;
• legal disputes (e.g., between neighbors).

These and other negative aspects must of course also be considered in an overall evaluation.

1.11 Conclusion

The results of such an evaluation may vary, depending on requirements, assessment and objectives. However, the positive aspects are always likely to prevail. The occasional inconvenience caused by trees should therefore be tolerated.

References

Arnberger, A. (2006). Recreation use of urban forests: an inter-area comparison. *Urban Forestry and Urban Greening* **4**, 135–144.

Bahamón, A. (ed) (2008). *Ultimate Landscape Design*. Te Neues, Kempen.

Bowler, D.E., Buyung-Ali, L., Knight, T.M. and Pullin, A.S. (2010). Urban greening to cool towns and cities: A systematic review of the empirical evidence. *Landscape and Urban Planning* **97**(3), 147–155.

Bucur, V. (2006). *Urban Forest Acoustics*. Springer, Berlin/Heidelberg.

Carreiro, M.M.; Song, Y.-C.; Wu, J. (eds) (2008). *Ecology, Planning, and Management of Urban Forests*. Springer, New York.

Clarke, J.H. (1996). *Dictionary of Practical Materia Medica*. Bd. 10, Grohmann, Bielefeld.

Coutts, M.P. and Grace, J. (1995). *Wind and Trees*. Cambridge University Press, Cambridge.

Cox, S. (2011). *Urban Trees – a Practical Management Guide*. Crowood Press, Marlborough.

Donovan, R., Hewitt, S.O., Owen, S. McKenzie, R. and Brett, H. (2011). *The Development of an Urban Tree Air Quality Score (UTAQS)*. Müller, Saarbrücken.

Hansmann, R., Hug, S.-M. and Seeland, K. (2007). Restoration and stress relief through physical activities in forests and parks. *Urban Forestry and Urban Greening* **6**, 213–225.

Hardin, P.J. and Jensen, R.R. (2007). The effect of urban leaf area on summertime urban surface kinetic temperatures: a Terre Haute case study. *Urban Forestry and Urban Greening* **6**, 63–72.

Harris, R.W., Clark, J.R. and Matheny, N.P. (2004). *Arboriculture*, Fourth Edition Pearson Education, Prentice Hall, New Jersey.

Heidt, V. and Neef, M. (2008). Beenfts to urban green space for improving urban climate. In: Carreiro, M.M.; Song, Y.-C.; Wu, J. (eds). *Ecology, Planning, and Management of Urban Forests*, pp. 84–96. Springer, New York.

Hofmann, M., Westermann, J.R., Kowarik, I., and van der Meer, E. (2012). Perceptions of parks and urban derelict land by landscape planners and residents. *Urban Forestry and Urban Greening* **11**(3), 303–312.

Konijnendijk, C.C. (2008). *The Forest and the City*. Springer, Berlin/Heidelberg.

Konijnendijk, C.C., Nilsson, K., Randrup, T.B. and Schipperijn, J. (2005). *Urban Forests and Trees*. Springer, Berlin/Heidelberg.

Kowarik, I. and Körner, S. (eds) (2005). *Wild Urban Woodlands*. Springer, Berlin/Heidelberg/New York.

Lee, A. and Maheswaran, R. (2011). The health benefits of urban green spaces: A review of the evidence. *Journal of Public Health* **33**(2), 212–222.

Lohr, V.I., Pearson-Mims, C.H., Tarnai, J. and Dillman, D.A. (2004). How urban residents rate and rank the benefits and problems associated with trees in cities. *Journal of Arboriculture* **30**(1), 28–35.

Matsuoka, R.H. and Kaplan, R. (2008). People needs in the urban landscape: Analysis of landscape and urban planning contributions. *Landscape and Urban Planning* **84**(1), 7–19.

McPherson, E.G. and Simpson, J.R. (2002). A comparison of municipal forest benefits and costs in Modesto and Santa Monica, California. *Urban Forestry and Urban Greening* **1**, 61–74.

Miller, R.W. (2007). *Urban Forestry*, 2nd edition. Waveland Press, Inc., Long Grov.

Nowak, D.J., Crane, D.E. and Stevens, J.C. (2006). Air pollution removal by urban trees and shrubs in the United States. *Urban Forestry and Urban Greening* **5**, 115–123.

Price, C. (2003). Quantifying the aesthetic benefits of urban forestry. *Urban Forestry and Urban Greening* **1**, 123–133.

Roloff, A. (2013). *Bäume in der Stadt*. Ulmer, Stuttgart.

Sanesi, G., Lafortezza, R., Bonnes, M. and Carrus, G. (2006). Comparison of two different approaches for assessing the psychological and social dimensions of green spaces. *Urban Forestry and Urban Greening* **5**, 121–129.

Schulz, H.-J. and Balder, H. (2000). The monetary value of street trees in cities as for example Berlin. *Mitteilungen aus der Biologischen Bundesanstalt für Land- und Forstwirtschaft* **370**, 302–311.

Secco, G., Zulian, G. (2008). Modeling the social benefits of urban parks for users. In: Carreiro, M. M.; Song, Y.-C.; Wu, J. (eds) (2008). *Ecology, Planning, and Management of Urban Forests*, pp. 312–335. Springer, New York.

Smardon, R. (1988). Perception and aesthetics of the urban-environment – review of the role of vegetation. *Landscape and Urban Planning* **15**(1–2), 85–106.

Stokes, J. and Rodger, D. (2004). *The Heritage Trees of Britain and Northern Ireland*. Constable, London.

Sugiyama, T.; Thompson, C.W. (2008). Associations between characteristics of neighbourhood open space and older people's walking. *Urban Forestry and Urban Greening* **7**, 41–51.

Tommasi, D., Miro, A., Higo, H.A. and Winston, M.L. (2004). Bee diversity and abundance in an urban setting. *Canadian Entomologist* **136**, 851–869.

Trowbridge, P.J. and Bassuk, N.L. (2004). *Trees in the Urban Landscape*. Wiley, Hoboken New Jersey.

Tyrväinen, L., Pauleit, S., Seeland, K. and de Vries, S. (2005). Benefits and uses of urban forests and trees. In: Konijnendijk *et al.* (eds). *Urban Forests and Trees*, pp. 81–114. Springer, Berlin/Heidelberg.

Velarde, M.D., Fry, G. and Tveit, M. (2007). Health effects of viewing landscapes – landscape types in environmental psychology. *Urban Forestry and Urban Greening* **6**, 199–212.

Yang, J., McBride, J., Zhou, J. and Sun, Z. (2005). The urban forest in Beijing and its role in air pollution reduction. *Urban Forestry and Urban Greening* **3**, 65–78.

CHAPTER 2

Urban trees: Features and requirements

Andreas Roloff

Technische Universität Dresden, Tharandt, Germany

2.1 Urban tree site categories

Urban trees are subject to several special stress factors, connected to and caused by living conditions that are extreme compared to forests (Figure 2.1). Of course, this varies greatly, depending on where in a city a tree is placed. Possible site types are ordered below, in Table 2.1, by decreasing naturalness and increasing stress.

Table 2.1 Urban tree sites, ordered by decreasing naturalness/increasing stress/artificial conditions.

- Urban woodlands and forests
- Parks, green spaces
- Private gardens
- Public squares and pedestrian precincts
- Road-side areas
- Roof terraces and roof gardens
- Covered atriums
- Buildings interiors

2.2 Special conditions for urban trees

Special conditions that must be considered for urban trees are listed in Table 2.2 (in alphabetical order).

2.3 Requirements and selection criteria

As a result of this, there are potential requirements and selection criteria for the nomination of species as urban trees. These are presented in Table 2.3 (in alphabetical order, with claim to completeness).

It may help a little that the mentioned criteria can be sorted into:
- aesthetic factors;
- aspects of design;

Urban Tree Management: For the Sustainable Development of Green Cities, First Edition. Edited by Andreas Roloff.
© 2016 John Wiley & Sons, Ltd. Published 2016 by John Wiley & Sons, Ltd.

Figure 2.1 Extreme living conditions for an urban roadside tree.

- ecological requirements of the species;
- tolerance of the species;
- positive effects on people;
- possible annoyance.

For more details on the species selection process and species suggestions and features, see Chapter 15 and:

- for temperate zone: Hightshoe (1988); Gilman (1996); Roloff *et al.* (2009, 2015); Berk (2015);
- for sub/tropical cities: Ping and Lynn (2001); Chin (2003); Johl and Bahga (2014).

The author's institute (www.tu-dresden.de/forstbotanik) has developed a free-access database called Citree for this purpose. Based on an extensive literature review, more than 390 woody plants were investigated to obtain a comprehensive assessment of specific characteristics by integrating 65 specific urban aspects. Within this study, a database was developed that allows users simultaneously to consider site characteristics, natural distribution, tree appearance, ecosystem services, management activities and the risks and interferences caused by urban woody plants. The user can choose from more than 60 criteria for species selection in order to choose the optimum species based on scientific findings (see Chapter 15). The Citree database is useful for preventing mistakes in planning that would otherwise result in high ecologic and economic costs. Choosing the right species for the right location may also increase the floristic biodiversity within tree planting.

Table 2.2 Special conditions for urban trees (in alphabetical order).

- All day sunshine/radiation
- Artificial light
- Damage/injuries (crown, trunk, root)
- Dog/human urine
- Drought stress
- Dust impact
- Emissions/pollution
- Increased temperature
- Leaks from pipes (e.g., gas)
- Nutrient deficiency
- Oxygen deficiency in soil/rooting area
- pH value neutral to alkaline (>7)
- Pruning/cutting
- Reflections from buildings
- Road salt impact
- Rooting space restrictions/limitations
- Single/solitary trees in focus
- Soil contamination (e.g., oil, cement, etc.)
- Soil disturbances
- Soil compaction
- Soil covering/sealing

Table 2.3 Requirements and selection criteria for species nomination (in alphabetical order).

Allergy risk (by pollen, hairs)

Bark: color, beauty, structure, thickness, drop of bark scales, etc., potential uses (coloring, bast), sunburn sensitivity

Benefits for animals (birds, mammals, insects) – nesting, protection, food

Costs of planting, maintenance

Crown size, shape

Crown transparency (species-specific)

Drought stress tolerance

Edible leaves, fruit, branches (also for animals)

Effects on the psyche

Features for bioengineering

Flowers: color, intensity, smell, time of flowering, annoyance (allergies, dust, etc.)

Fruits: number, size, color, time of ripeness, edible, nutrient content (also for animals), annoyance (also caused by animals), dropping period/time

Growth rate of young/mature trees

Emission tolerance (SO_2, NO_x, O_3, HF, etc.)

Emission reduction potential (fixation/absorption)

Invasiveness/spreading potential: problems

Late frost susceptibility

Leaf coloration: intensity, period/time

Leaves of evergreen/deciduous, leaves stay brown on branches in autumn

(continued)

Table 2.3 (continued)

Leaves: shape, size, color, time of flush/autumn coloration/leaf drop, potential uses (decoration, food), annoyance (leaf drop, shading), self cleaning/dust accumulation

Life expectancy

Light requirements (young, mature plant)

Maintenance requirement/costs

Mycorrhiza necessary

Nature conservation aspects (indigenous, introduced, rare species, tree for birds or other animals, biotope, habitat)

Neighborhood situation (leaf drop, fruit drop, shading, hanging branches)

"New" species (test, experiment)

Nitrogen fixation by roots (symbiosis with bacteria)

Noise reduction potential

Pathogens/pests (current, future)

Personal relationships/experience with specific species (childhood, life stages, house/family tree, etc.)

pH value amplitude

Pioneer features

Toxicity (leaves, fruit, branches)

Potential for dust fixation/reduction

Pruning/cutting tolerance

Psychological effects

Radiation tolerance (single trees)

Risk of breakage (branches, trunk)

Root suckers

Rooting intensity

Rooting type (flat, heart, tap root)

Salt tolerance

Shading

Size of mature tree (small, large)

Slow growth

Soil requirements (water, nutrients, pH value, etc.)

Soil tolerance: drought, contamination, wetness, flooding, compaction, pH value, etc.

Spring sap

Sunburn sensitivity (bark, leaves)

Thorns, spines

Trunk dominance (tendency to coppicing/root suckers)

Use of firewood

Use of wood (furniture), bark (dye), leaves (decoration), flowers (recipe), fruit (food), spring sap, etc.

Vegetative spread with root suckers, branch layering

Visual barrier effect

Water demand

Wind protection

Wind throw risk

Winter frost hardiness

2.4 Conclusions

The detailed enumeration of urban tree selection criteria shows that the number of factors that need to be considered is impressive – perhaps even intimidating to some. The intention of this list is to help identify those factors that are relevant for the specific situation, and to use these for selecting the species.

References

Berk, G. v.d., (2015). *Trees*, 2nd edition. Boomkwekerij Gebr. van den Berk, Sint-Oedenrode/NL.

Chin, W.Y. (2003). *Tropical Trees and Shrubs – a Selection for Urban Plantings*. Sun Tree Publishers, Singapore.

Gilman, E.F. (1996). *Trees for the Urban and Suburban Landscapes*. Delmar Publishers, Albany NY.

Hightshoe, G.L. (1988). *Native Trees, Shrubs, and Vines for Urban and Rural America*. Nostrand Reinhold, New York.

Johl, H.S. and Bahga, S.S. (2014). *Trees in Urban Habitat*. Lightning Source, Milton Keynes, UK.

Ping, T.S. and Lynn, W.M. (2001). *Trees of Our Garden City*. National Parks Board, Singapore.

Plotnik, A. (2000). *The Urban Tree Book*. Three Rivers Press, New York.

Roloff, A., Korn, S. and Gillner, S. (2009). The Climate Species-Matrix to select tree species for urban habitats considering climate change. *Urban Forestry and Urban Greening* **8**, 295–308.

Roloff, A., Weisgerber, H., Lang, U.M. and Stimm, B. (eds.) (2015). *Enzyklopädie der Holzgewächse*. Wiley-VCH, Weinheim.

CHAPTER 3

Fundamentals of tree biology for urban trees

Doris Krabel

Technische Universität Dresden, Tharandt, Germany

3.1 Morphological and anatomical features

3.1.1 Trunk

In comparison to woody shrubs, trees are characterized by a dominating trunk, whose lowermost part can reach a great height with age, and which is generally free of branches. In contrast, the shape of shrubs is characterized by several trunks, without a dominating stem, a height of up to ten meters, and the development of strong shoots at the basal part of the plant. Shrubs tend to rejuvenate from the bottom (Roloff, 2013). The transverse sections of a woody shoot present three major parts (Figure 3.1).

The outermost part is the bark, followed by the secondary phloem (bast or inner bark), which is separated by a thin layer of cambial cells from the wood (secondary xylem), which comprises most of the volume of the trunk. From an anatomical viewpoint, woods can generally be classified into two main groups – softwoods and hardwoods. The term "softwood" is applied to gymnosperm wood, and "hardwood" to dicotyledon wood (Evert, 2006).

Gymnosperm wood, to which conifer wood belongs, is more homogenous in structure, whereas dicotyledon wood is more heterogeneous, due to the higher number of different cell types. The secondary xylem (wood) of conifers is mainly composed of elongated cells (tracheids) of average length 2–5 millimeters and wedge-shaped at the end. These cells are connected with bordered pits (Evert, 2006).

Due to the seasonal cambial activity (section 3.2.3), early and late wood tracheids can be separated. Early wood cells are formed in spring and summer. In microscopical cross-sections, they show a larger diameter and thin cell walls, and the bordered pits have a circular shape. Late wood tracheids (fiber tracheids) are characterized by increased cell wall thickness, more oval-shaped pits and a smaller diameter.

Early and late wood cells formed within one growth period comprise the annual growth ring, and occur in deciduous and evergreen trees. Growth rings are not only typical for wood from the northern or southern temperate zones, with dormant and growth activities, but also for regions in the tropics, where seasonality of growth is triggered by severe dry and humid periods. In these regions (e.g., Australia, Amazonia), most trees lose their leaves during the dry season and produce new ones shortly after or during the rainy season. During this time, xylem cells are produced and form a growth ring (Evert, 2006).

Urban Tree Management: For the Sustainable Development of Green Cities, First Edition. Edited by Andreas Roloff.
© 2016 John Wiley & Sons, Ltd. Published 2016 by John Wiley & Sons, Ltd.

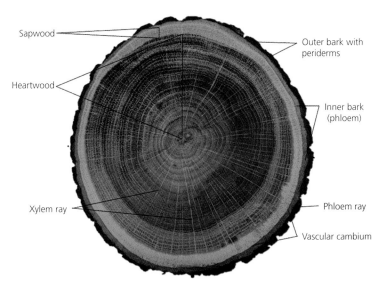

Sapwood

Outer bark with periderms

Heartwood

Inner bark (phloem)

Xylem ray

Phloem ray

Vascular cambium

Figure 3.1 *Quercus* stem, showing the transverse surface. (Photograph by I. Herzog.)

The width of individual growth rings may vary from growth period to growth period according to environmental factors such as rainfall, temperature and/or insect attacks.

Besides the water-conducting and stabilizing cells, ray cells are also present in conifer wood. The rays are composed of parenchyma cells, whose function is the storage of assimilates and ray tracheids which conduct water in a radial direction. Furthermore, resin ducts characterize the xylem of most conifers in the radial and longitudinal directions. Axial storage cells (parenchyma cells) are sometimes present in conifer wood (e.g., in *Taxodiaceae* and *Cupressaceae*).

Dicotyledon wood is much more complex than that of conifers. Except for some primitive dicotyledons which have no vessels, the xylem of most dicotyledonous species have two kinds of cells for water transport (vessels and tracheids), one or more categories of fibers (stabilizing cells), axial storage cells (axial parenchyma cells) and rays of one or more kinds (Evert, 2006). The vessels especially characterize dicotyledonous wood. These structures consist of cells of large diameter (pores), whose transverse walls are perforated or, at most, reduced, so that they may form a tube of several meters and serve as the most effective water transport system from the roots to the crown.

Regarding the distribution of vessels within a growth ring, two principles can be recognized: diffuse porous orientation of vessels more or less of the same size, and distribution throughout the growth ring (e.g., *Acer saccharinum, Acer saccharum, Betula nigra, Cornus florida, Fagus grandifolia, Liquidambar styraciflua, Liriodendron tulipefera, Platanus occidentalis, Tilia americana, Umbellularia californica*); a and ring-porous structure, with larger pores in the early wood than in the late wood (e.g., *Carya pecan, Castanea dentata, Catalpa speciosa, Celtis occidentalis, Gleditsia triacanthos, Morus rubra, Paulownia tomentosa, Quercus spp., Sassafras albidum, Ulmus americana*) (Evert, 2006).

Typically, with age, the inner part of the wood loses the functions of water conductivity and storage. This process is often associated with color and odor changes caused by the infiltration of the wood by various substances such as oils, gums, resins and tannins. However, the outer part of the xylem still contains living cells and reserve materials. The first wood is called heartwood, and this is generally darker and heavier than the still-living sapwood. The proportion of sapwood to heartwood, and the degree of differences, varies from species to species (Evert and Eichhorn, 2013).

During the annual vegetation cycle, early and late wood are formed, and these comprise the annual growth ring. The formation and dimensions of annual growth rings depend on internal and numerous environmental factors. As these factors vary from growth period to growth period, annual growth rings are suitable indicators to deduce changes in the trees' environment (this is the basis of dendrochronological research). Palm trees, and the majority of trees which do not grow between 25° and 65° latitude, are not suitable for tracing growth rings (Firestone, 2003), because either these species do not form growth rings (palm trees) or the borders between different growth rings cannot be separated precisely, or they form several growth ring-like structures within one vegetation period.

Reaction wood is formed by a branch or stem as a reaction to having to counteract the force of gravity or mechanical stress like wind (Evert and Eichhorn, 2013). In angiosperms, reaction wood is formed as tension wood on the upper side of the leaning part, whereas compression wood is formed by conifers on the underside of the bended part. Reaction wood can be recognized by an eccentric annual growth ring, which is caused by unilaterally higher cambial activity.

Palm trees are among those species that do not belong to conifers or broadleaved trees, but instead belong to the monocotyledonous group of plants (e.g., bamboo). Their wood consists of primary vascular bundles (composed of cells that regulate water and assimilate transport) embedded in a parenchymatous ground tissue, which means that palm "wood" is primary tissue, and is not differentiated by a vascular cambium (which forms the secondary xylem), which is missing in palm trees. Consequently, the trunks of palm trees do not show structures like the abovementioned growth rings. Even though their stem does not consist of secondary xylem like that of conifers and broadleaved trees, mature stems may grow up to 50 meters in height, such as the Andean wax palm, and reach dimensions of 1 meter across (Parthasarathy and Klotz, 1976). The texture and hardness of a palm stem depends on the distribution of the vascular bundles and the amount of sclerenchyma (cells for stabilization).

3.1.2 Roots

The major functions of roots are: stabilization and anchorage of the tree to the ground; absorption and conduction of water and inorganic nutrients; storage of carbohydrates and production of phytohormones; and, in part, vegetative reproduction. Due to these diverse functions, the root tissues appear anatomically quite different.

The first root (radicle) of a plant descends from the embryo. It grows downward, due to gravitropism. In the development that follows, this root produces at the tip (apical meristem) a primary root (taproot), which branches out in lateral roots (secondary roots). Together, these form the root system.

In monocotyledons (palm trees), the first root commonly lives only a short time, and the major root system is formed by adventitious roots arising on the shoot. Although these roots also become branched, they form a more homogenous root system compared to the taproot system (Esau, 1977).

Generally, the oldest roots are in the vicinity of the root collar, while the youngest ones are near the root tip. Before a taproot increases in thickness (induced by secondary growth), it shows a primary organization.

The morphological organization of a root is characterized by the rootcap (calyptra) at the tip. The rootcap protects the delicate apical meristem against mechanical damage, and assists the growing root in penetrating the soil. Generally, those root tips are coated with a more or less large amount of mucilage, which promotes the adherence of soil particles to the root tips and root hairs (Evert, 2006). From the apical meristems, new cells are produced continuously, while old ones are sloughed off, which results in the root growing at the tip. The region of active cell division, which is the youngest part of the root, is followed by a region of cell enlargement which overlaps with the region of cell differentiation. The latter is the region where the cells fulfill their functions. The absorption of water and soluble inorganic nutrients occurs in adjacent root hairy zone, where the cells are already fully differentiated. The total length from root cap to root hairy zone usually comprises less than one centimeter. Nevertheless, most of the water is absorbed by this part of the root, which is why a large number of fine roots is essential for favorable development of the plant.

Cross-sections of young roots show delicate cells which encircle the root tissue inside. This tissue is called the epidermis (synonym rhizodermis) and, in contrast to the epidermis of leaves, it is not covered with a transpiration-reducing waxy layer (cuticula). Instead, these epidermis cells are specialized as an absorbing tissue that bears root hairs which extend the absorbing surface of the root. The life span of root hairs is limited to generally only a few days.

Similar to the stem, roots can increase in diameter. When extending in thickness, the epidermis collapses and an exodermis occurs beneath the epidermis and becomes a new protective layer. Usually, exodermis cells have a suberin lamella covered by a cellulose wall (Evert, 2006). This layer prevents absorption of water, so that the main task of the exodermis is protection of the root against mechanical damage and avoidance of desiccation. More central, the cross-section shows the cortex, which comprises the largest part of the root cross-section. These are mostly parenchyma cells which store assimilates (e.g., starch) and, due to a comparatively loose arrangement of cells, the tissue provides the central part (vascular cylinder) of the root with oxygen. Between the cortex and vascular cylinder, two additional tissues are located. The more peripheral tissue, the endodermis, is comprised of tightly arranged cells which control the transport of soil solution into the vascular cylinder. The cell layer which covers the periphery of the vascular cylinder is the pericycle. The pericycle is commonly one cell layer in thickness, and it is responsible for lateral root development and development of a vascular root cambium.

Generally, three different morphological types of root systems can be distinguished: tap-root system (e.g., *Larix europaea, Quercus petraea/robur, Ulmus* spec.); heart-like root system (e.g., *Betula* spec., *Tilia* spec., *Acer* spec., *Fagus sylvatica*); and shallow root system (e.g., *Fraxinus excelsior, Picea abies, Populus tremula, Sorbus aucuparia*) (Roloff, 2013) (Figure 3.2).

The formation of the architecture of the root systems is genetically determined, but varies with age and site (see Chapter 4), so a clear classification to one of the root systems is mostly only possible for young trees. The distribution of roots horizontally and vertically in older trees differs between tree species but, even within one species, the shape of the root architecture may also differ due to soil composition, groundwater level and nutrient availability. Thus, rather shallow roots may develop due to a high groundwater level (Figure 3.3).

In some cases, adventitious roots originate from the stem, branch or other "unusual" parts of root development (Evert and Eichhorn, 2013). Such adventitious roots are common for monocots, but may occur also in dicots under certain circumstances, such as damage, inner rot, or pile-up of the stem or root formation at cuttings.

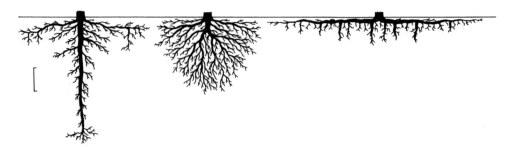

Figure 3.2 Scheme showing the main types of root architecture (left to right): tap-root system, heart-like root system and shallow root system (Figure: A. Roloff).

Figure 3.3 80 year old *Picea abies* in an urban forest uprooted by storm. The shape of the horizontally oriented roots can be reconstructed in the soil.

3.1.3 Mycorrhizae

Fungi may play an important role in enhanced water and nutrient uptake of roots (see Chapter 10, section 10.3.7). Experiments with seedlings cultivated under sterile soil and environmental conditions show that the seedlings exhibit reduced and poor growth, or that some die due to malnutrition (Evert, 2006), even when the soil contains ample nutrients. However, if some forest soil containing fungi is added to the substrate, the plants show normal growth behavior.

Optimal growth seems to be possible if the trees' roots are associated with a specific fungus. Such association is called mycorrhiza (plural mycorrhizae), and it is generally a symbiotic association of fungi and roots of vascular plants. Mycorrhizae are classified into two principal forms, according to the relationship between the fungus and the cortical cells of the host (Esau, 1977). The ectomycorrhiza is accompanied by anatomical changes of the feeder root tip, which loses the root hairs due to invasion of the cortex cells by the fungus. The root tip swells and maintains a characteristic shape. The endomycorrhizal (vesicular-arbuscular or VA mycorrhizae) roots are similar to uninfected roots, but show a darker color. The tight anatomical association of both partners allows the exchange of specific nutrients. The host makes sugars, amino acids and other organic substances available for the fungus, while the fungus converts minerals from the soil and decayes organic material and makes these substances accessible to the tree (Esau, 1977).

Most ectomycorrhizal associations in trees are caused by basidiomycetes in the mushroom group (*Agaricales*) or the puffball group (*Gasteromycetes*). Some of the tree species that are associated include *Abies, Alnus, Betula, Carpinus, Castanea, Castanopsis, Corylus, Cyclobalanopsis, Engelhardtia, Eucalyptus, Fagus, Gilbertiodendron, Larix, Lithocarpus, Nothofagus, Pinus, Pseudotsuga, Quercus, Shorea, Tetraberlinia and Tsuga* (Agerer, 1997) while endomycorrhizal associations are formed with *Cordyla, Daniellia, Detarium, Erythrophleum* (Thoen and Bâ, 1989), *Acer, Camellia, Cornus, Fraxinus, Juglans, Liquidambar, Malus, Persea, Prunus, Rhododendron, palms* and others. Fungi involved in VA mycorrhizae are generally phycomycetous fungi belonging to the genus *Endogone* (Blanchard, 2012).

Due to the generally positive impact of mycorrhizae on tree growth and vitality, a number of "mycorrhiza products" have been commercially developed. When planting in urban areas for the first time, these products can be selectively incorporated into the soil, thus supplying the trees with their specific mycorrhiza partner (see Chapter 10).

3.1.4 Secondary growth

Most of the volume in woody plant species such as trees is formed by a meristematic tissue, which is located between the secondary xylem (synonym: wood, corresponds to the solid lignified inner part of the stem) and the outer part of the stem, which is the inner bark (synonym: secondary phloem, secondary bark). These meristem cells, which remain able to divide during the whole lifespan of the tree (sometimes several hundred years) surround roots, trunk and branches in a more or less cylindrical sheath called the vascular cambium or cambial meristem. The cambium is responsible for the secondary growth (increase of diameter) of a tree or shrub and, hence, the production of wood (Figure 3.4). In monocots, like palm trees, and herbaceous plants (e.g., herbs), such a meristematic sheath is missing, and growth of a specific organ is terminated by the differentiation of primary tissue. In the case of palm trees, the stability and radial growth of the trunk is enabled by primary growth of tissue located near by the shoot tip and sclerenchymatic cells inside the trunk.

The specific characteristic of this meristematic sheath is that it is an extremely flexible tissue, composed of cells which divide and differentiate into a number of different types of cells (derivatives) with completely different functions. These derivatives are clearly different in morphological, anatomical and physiological performance (Figure 3.5).

By this means, the cambium produces wood (xylem) cells and phloem cells. Xylem cells are mainly responsible for the transport of water and soluble nutrients, whereas phloem cells, located in a comparably thin layer at the periphery of the trunk, mainly conduct (sieve cells, sieve elements) or store assimilates (parnechyma cells), or serve as stabilizing elements (fibers). The radial elements of this part of the trunk (phloem rays), act like wood rays, for stabilization and the supply of the inner trunk with assimilates and soluble nutrients.

Figure 3.4 Trees can grow tall (*Acer pseudoplatanus*) because their stem, roots and branches increase due to secondary growth. The main portion of produced cells is comprised of xylem (wood) cells, which supply the trunk and crown with water but also give stability to the organs. (Photograph by H. Merten.)

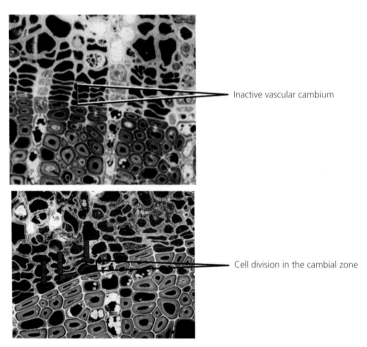

Inactive vascular cambium

Cell division in the cambial zone

Figure 3.5 Vascular cambial meristem of the *Fagus sylvatica* tree. Various cell types are differentiated by the cambium. Outwards, cambial cells differentiate essentially cells of assimilate conduction and storage (sieve elements, companion cells, parenchyma cells, phloem rays), as well as stability (phloem fibers). Inwards, it produces cells of water transport (tracheids, vessels, xylem rays), carbohydrate storage (parenchyma cells) and stability (xylem fibers – see thick cell walls) (×300).

Compared with deciduous trees, the wood structure of conifers appears much more homogenous, because the number of different cell types is smaller than in broad-leaved trees.

According to the seasons, we find fundamental differences in cell division activity in the cambial zone in the temperate areas. Usually, cell division activity decreases in autumn and stops completely in winter. Despite cold temperatures, day length is one environmental factor which triggers the entrance of meristematic cells into this period (Wareing, 1956; Denne, 1974), called "dormancy". During this phase, which lasts in northern temperate regions roughly from October until March, dormancy is not of consistent intensity. After around two to four weeks of absolute inactivity, the cambial cell division process can be reactivated by light treatment (Mellerowicz *et al.*, 1992), which is not possible during the preceding weeks.

Under natural conditions, the reactivation of cambial cells restarts in spring. This procedure is accompanied by a morphological transformation of the cells. They begin to swell and the cell walls become less rigid. In parallel, the meristem absorbs water, and stored assimilates are dissolved. A visible sign of this transformation is the easy separation of the bark from the wood, which is called "slipping of the bark" (Esau, 1977).

At this time, the cells are less frost-resistant than during cambial inactivity (Ladefoged, 1952). During the phase of transformation into a meristem, which is highly active in cell division processes, huge vacuoles encompass the total cell, rather than the several small vacuoles that characterize cambial cells. At the period of intensive cell division, the meristem is specifically able to regenerate and to perform wound reactions (see section 3.2.4). Microscopical investigation have shown that, during this phase, up to 20 new cells per day were differentiated from one single cambial cell (Eschrich, 1995).

In deciduous trees, reactivation processes in spring or after drought seasons (sub-tropics) seem to be correlated with bud development (e.g., Sachs, 1981; Evert, 1987; Krabel, 1998). At this time, however, details of this complex procedure are not fully understood.

3.1.5 Periderm and bark

Periderm is defined as a secondary multi-layered tissue at the surface of the stem, branches or roots of woody plants. It protects severe tissue, such as the vascular cambium, from mechanical damage, intense light, exsiccation, extreme temperatures or damage by pathogens and insects.

Triggered by a physical pulse (pressure or tension), which is provoked by the secondary growth, cells belonging to secondary phloem start to divide and form a cork cambium (phellogen). This replaces the original primary epidermis cells, whose cells start to stretch, flatten and burst, and finally these cells die off. Similar to the vascular cambium, the phellogen is usually composed of one or few cell layers of meristematic cells, and divides periclinally to the outside, producing the phellem or cork cells and phelloderm cells to the inside (Figure 3.6). The phellem especially, whose cells contain suberin, protect the inner stem against desiccation and pathogens.

Compared to the thickness of the phellem, phelloderm consists only of few cell layers. The total tissue which is formed in such a way, and which consists of phellogen (cork cambium), phellem (cork) and phelloderm, is called periderm.

As a consequence of additional secondary growth, the stem increases in circumference and new phellogen cells are differentiated. These contribute to a new, younger periderm. In such a way, new protective periderm tissues are continuously produced. The old ones at the periphery of the plants organ die off and, in some cases, these parts are "pushed off".

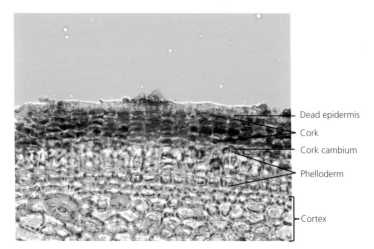

Dead epidermis

Cork

Cork cambium

Phelloderm

Cortex

Figure 3.6 *Sambucus nigra*: Arrangement of a typical periderm (×50).

Such a procedure can be observed in summer on plane trees. In this case, however, old periderm cells and secondary phloem cells, which are localized between two periderms, dry out because they are cut off from assimilate support.

The bark of a woody plant or shrub is the outermost part of the circumference, which is located between the youngest tissue (the innermost), the newly differentiated periderms and the oldest outermost lying periderm layers. This also includes secondary phloem cells, which are located in between the different periderm layers.

The varying shape of bark morphology of trees and shrubs is related to differing arrangements of the various periderms. The plane tree, for example, shows a bark with squarrose shape, which indicates that phellogen cells are arranged in a specific area. Other tree species, such as birch or cherry, show phellem layers which are arranged in stripes/rings around the tree's circumference, and this is called ring-cork. To be anatomically correct, this is not "ring-bark", as these are not bark layers but layers of phellem (Bartels, 1993).

Different types of bark are not only related to a different organization of the periderms, but also to the color and thickness of cork. Thus, for example, the cork of *Quercus suber* can be harvested every 9–12 years to produce cork for commerce.

Within the periderm, we can often find wart-shaped structures at the surface of young twigs or stems. These structures are living cells called "lenticels". They are responsible for gas exchange, and therefore maintain the contact between the atmosphere and the tissues inside the stem. Investigations with radioactive carbon dioxide showed that, in the vicinity of lenticels, green bark cells are photosynthetically active. This indicates that, besides leaves, stems and twigs are also able to produce assimilates (Langenfeld-Heyser, 1989).

3.2 Tree growth and growth reactions

3.2.1 Photosynthesis – the fundamental growth process

Photosynthesis (literally: synthesis from light) is the metabolic process which enables life on our planet, because this process captures light from the sun and uses it for the transformation of carbon dioxide (CO_2) and water (H_2O) into carbohydrates ($C_6H_{12}O_6$ and others) and the release of oxygen (O_2). Under favorable environmental conditions (sufficient light, water, CO_2 and nutrients), photosynthetic reactions take place, mainly in leaves, but also in other green organs, triggered by chlorophyll molecules within a plant (e.g., bark of young shoots of ash or lime tree).

The total process can be divided into two major reactions: light reactions, or energy transduction reactions, during which light energy is converted to chemical energy, while water is consumed and oxygen produced; and carbon-fixation reactions (or dark reaction), in which carbon dioxide from ambient air is transformed into organic compounds (assimilates). The latter reaction does not need light directly, but it does need the products (energy-rich molecules) which are produced during the light reactions. The assimilates that are produced in such a manner can be converted to sugars (e.g., sucrose) and transported from the leaves ("sources") to carbohydrate-consuming locations (growing tissue, "sinks"), or can be stored as starch molecules and hydrolyzed on demand. Only a small proportion of photosynthetically produced carbohydrates serve the plant for biomass aggregation. The larger proportion of those products is used for keeping the plant's metabolic activities running.

Respiration is another important biosynthetic process, in which carbohydrates are oxidized into carbon dioxide while energy rich molecules are produced. Respiration is considered to start generally with glucose, an end product of the hydrolysis of sucrose and starch. In the oxidation process, the glucose is split apart and water is released by the combination of hydrogen atoms and oxygen. In contrast to photosynthesis, respiration is independent from sunlight, which means that this process takes place not only in leaves but also in organs missing chlorophyll (e.g., roots), and also at night or during winter, when light intensity is low. If the balance of photosynthesis and respiration is positive, then dry mass can be investigated into growth processes.

3.2.2 The role of water

A plant body mainly consists of water, sometimes comprising more than 90% of its fresh weight. Water is a fluid for chemical reactions, a base product of photosynthesis, a solvent for chemical reactions, and it transports organic and inorganic nutrients from one organ to the other. Beside these important functions, water keeps plant cells under pressure (turgor) and gives non-lignified tissues stability.

In order to fulfill these functions, water has to be taken up from the soil by the root system and transported to the crown and other compartments where water is needed. In trees, the distance between the root system (the location for water uptake into the plant body) and the leaves (the locations for water demand and release) might be quite large (more than 100 meters in the tallest trees). Bridging over such large vertical distances requires an efficient system of water movement and conductivity (Figure 3.7).

Water enters the plant through the younger parts of the roots, where it is absorbed by the enormous surface area of the root hairs. From the root hairs, it moves through the cortex, the endodermis and into the vascular cylinder, respectively – the water conducting elements of the xylem.

The driving force for this movement across the root is the difference in water potential between the soil solution at the surface of the root and the water inside the xylem cells (xylem sap). The driving force of water flow from the roots in vertical direction to the crown is the transpiration stream.

Evert and Eichhorn (2013) report that the loss of water in a single day by a single tree growing in a deciduous forest of south western North Carolina is between 200–400 liters (50–100 gallons). This immense loss of water vapor, known as transpiration, is mainly

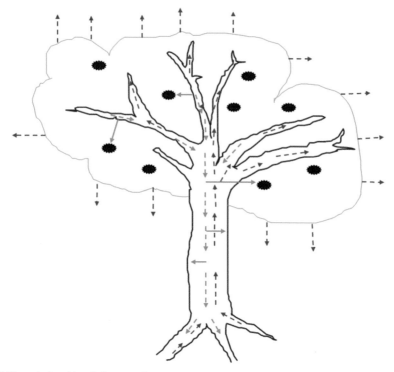

Figure 3.7 The relationship of photosynthesis to transpiration. Assimilates (green arrows) are produced by photosynthesis and transported to locations of demand (e.g., fruits, red dots). Transpiration stream (blue arrows); water is taken up by the roots and mainly released by stomata of the leaves.

carried out by the leaves, as the most important organs of transpiration, but flowers and shoots also emit water via the surface. Leaves are equipped with numerous little pores (stomata). This is a complex of cells composed of guard cells and stomata accessory cells, which open and close depending on a number of environmental but also internal factors, in order to regulate the carbon dioxide uptake from ambient air.

Following a concentration gradient, carbon dioxide enters the leaf cells via open stomata, but this implies that water, wherever it is exposed to unsaturated air, evaporates. The need for water in such large amounts is a dilemma, because carbon dioxide has to be taken up for photosynthesis, but transpiration can be regarded as a necessary cost for photosynthesis. The whole process of water movement from the soil into the plant and the atmosphere is only possible because the leaves are equipped with stomata, and the plant body with a system of cells specialized for water transport.

3.2.3 Seasonal dynamics

As trees are the longest living organisms we know on our planet, and as their environment is comparably quickly changing, they have to be adaptable in every sense (see Chapter 16). One possibility of individual adaptation is the adjustment of the metabolism as a response to changing environmental conditions (part of the phenotypic plasticity of the individual). In temperate zones, a well-ordered accumulation of assimilates and related compounds are built up during the growing season, mostly stored in autumn and winter and hydrolyzed in spring. In parallel, this process is accompanied by cambial cell production and increased metabolic activity respectively, reduced metabolic activity and other obviously visible growth processes such as bud burst and early leaf expansion (Nunez *et al.*, 2010).

In deciduous trees from tropical areas with dry and rainy periods, Barajas-Morales *et al.* (1997) found a decrease in starch at the beginning of the rainy season, coinciding with the growth of shoots. In particular, the metabolism of the vascular cambium, as thin layer of omnipotent cells, is closely adapted to environmental changes like changing temperatures, seasonal fluctuations, drought, and so on. Calamities (e.g., insect or pathogen attacks) or damage cause a more or less intense reaction of the cambial cells, depending on the status of activity (see section 3.1.4).

3.2.4 Wound reactions

Damage in trees is a portal of entry for microorganisms which induce discoloration and decay and, in numerous cases, they might decrease the vitality of the plant. In each case, the tree reacts, with compartmentalization of the damaged tissue in order to protect the intact tissue against invasion of pathogens (fungi and bacteria – see Chapter 6). The concept "CODIT" (Compartmentalization of Decay in Trees) has been developed by Shigo and Marx (1977) and Shigo (1979), and explains the different chemical, anatomical and physiological processes of responding on wounds.

Referring to Shigo and Marx (1977), the trees' reaction is focused on four compartments, which these authors call "walls". "Wall 1" restricts the invasion of pathogens in axial direction; "wall 2" builds a radial barrier; and "wall 3" diminishes tangential invasion of microorganisms. These compartments are predetermined by the anatomical structures of the xylem and, thus, the effectiveness of the compartmentalization increases from "wall 1" to "wall 4":

- "Wall 1" is the weakest barrier, because the water-conducting xylem elements (vessels and tracheids) are organized in a longitudinal direction in part over several centimeters, with the consequence that pathogen invasion may take place over quite a long distance inside these elements.
- "Wall 2" is more effective, due to the radial organization of growth rings and within each growth ring, especially late wood cells with small lumina and thicker cell walls, which form an intensive barrier (Schwarze, 2001).
- Ray cells form "wall 3" which seals better off than "wall 2" because, tangentially, a number of cell walls and a low oxygen concentration prevent microorganisms from invading. The three boundaries form a reaction zone around the damaged tissue that is mostly discolored and consisting of necrotic cells.
- Finally, "wall 4" (barrier zone), which is the strongest barrier, is composed of living cambial cells. The vascular cambium forms a barrier between the damaged xylem cells and newly developed cells (Figure 3.8).

The shape of discoloration, especially in axial direction of the stem, differs with the tree species. *Betula* and *Picea*, for example, show the most intense discoloration in the central part of the stem, while *Fraxinus* has the largest area of discoloration next to the cambium. In *Fagus* and *Acer*, a more circular discoloration around the wound can be observed (Dujesiefken *et al.*, 1991).

Based on the concept of Shigo and Marx (1977), Dujesiefken and Liese (2006) suggest a model which explains the complex wound reactions of a tree, not with anatomical structures but with four levels of defense. These steps encompass:

1. Reaction of the tree to air infiltration and desiccation of the disrupted tissue.
2. Reaction to the invading pathogen.

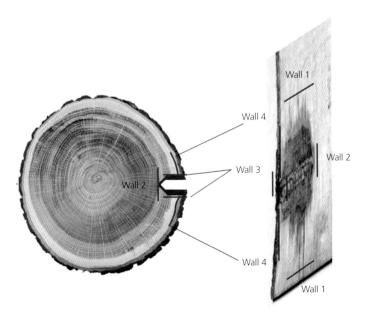

Figure 3.8 CODIT-Model. Effectiveness of compartmentalization increases from "wall 1" to "wall 4". (Photograph by I. Herzog.)

3. Consequences of the distribution of the pathogens within wood cells.
4. Successful encapsulation of the microorganisms and compartmentalization.
Therefore, Dujesiefken and Liese suggest the term "CODIT-Principle".

At the cellular level, protection against air embolism and invading microorganisms is achieved by the plugging of water-conducting elements, by the secretion of antimicrobial substances and by differentiation of new tissues (wound tissue), which compartmentalizes the destroyed tissue. Wound tissue formation generally starts with the production of thin primary walled parenchyma cells (callus) at the axial edges of the wound (lateral callus). Based on these young callus cells, a new vascular cambium is reformed and, later on, wound phloem and wound xylem cells develop. This tissue differs from regular xylem, in so far as the fibers are modified with thickened cell walls and increased lignin content (Stobbe *et al.*, 2002) (Figure 3.9). This total process is comparable to repair work on damaged areas at the tree's body, and not to a healing process (Shigo, 1979).

Besides the genetic constitution of the tree (see Chapter 16), a number of factors influence the dimensions of a wound reaction. Effective and quick-reacting tree species include *Fagus sylvatica, Quercus spec., Tilia spec., Pseudotsuga menziesii, Acer spec, Taxus bacata and Pinus spec.*, whereas a weak compartmentalization is described for the genera *Betula, Alnus, Fraxinus, Populus, Aesculus, Salix and Picea* (Dujesiefken *et al.*, 1991; Oven *et al.*, 2000; Jasulka and Blatt, 2001).

The variability of such a reaction is based, for example, for broad leaves in the area of the water-conducting cells, the proportion of parenchyma cells (Dujesiefken and Liese, 2006; see Chapter 11), climate and temperature, and the vitality and age of the trees. More sound trees show an increased response than less healthy trees. This is also true for conifers but, in this case, the production of resin is essential for the degree of forming boundaries. In addition, the effectiveness of a wound reaction depends on the depth of the wound. With increased age of the annual growth ring (the more central growth rings), the

Figure 3.9 Lateral wound callus formation (three years old) on *Tilia platyphyllos* stem as a consequence of severe damage. (Photograph by S. Herzog.)

proportion of living cells which might trigger the wound reaction decreases, so that only few cells are left to contribute to the compartmentalization process (Kowohl *et al.*, 2001).

A wound at the surface of the stem, without touching the vascular cambium, does not usually induce a reaction in the xylem. As soon as the vascular cambium and subjacent tissue are involved in the deletion, a reaction, as described above, will be induced.

In numerous cases, not only parenchyma cells start to form wound tissue, but also living cells from xylem rays. By this means, the area of the wound or part of the wound will be covered by undifferentiated cells (surface callus). Later on, these cells may divide and differentiate into wound phloem, wound cambium and wound xylem (Dujesiefken, 2001; Dujesiefken and Liese, 2006). The extent of a wound does not seem to influence the intensity of the wound reaction. Even small injuries, such as caused by a needle, induce a pronounced effect of compartmentalization (Dujesiefken, 2001; Jaskula and Blatt, 2001; Kersten, 2001; Kowohl *et al.*, 2001).

3.3 Conclusions

Trees are long-living organisms that are well adapted to various sites on our planet. Due to their complex genetic constitution and a number of morphological, anatomical and physiological features, they are able to withstand drought, flood, frost, heat and other environmental influences over decades, and with changing intensity. One such feature is the vascular cambium, which enables this group of organisms to produce immense amounts of biomass but which, on the other hand, enables wound reactions to repair small and large damage, or enables reaction wood to compensate for mechanical influences on the tree. Beside this, cambial cells trigger physiological processes related to seasonal changes or vegetative reproduction.

References

Agerer, R. (1997). *Colour Atlas of Ectomycorrhizae*. Einhorn-Verlag Eduard Dietenberger GmbH, Schwäbisch-Gmünd, Germany.

Albrecht, S. (2005). *Eignung der in der Baumpflege benutzten Schraube Baum-TagII der Firma d.b.g. für den Forstbetrieb unter besonderer Berücksichtigung der Wundreaktion von* Pinus sylvestris, Pseudotsuga menziesii, Quercus robur, Fagus sylvatica *und* Robinia pseudoaccacia *auf diese Schraube*; Diploma thesis, Dresden University of Technology.

Barajas-Morales, J., Pérez-Jiménez, L.A. and Chiang, F. (1997). Seasonal fluctuaciones of starch in wood and bark of trees from a deciduous tropical forest in Mexico. *Los Anales del Instituto de Biología, UNAM (Serie Botánica)* **68**, 7–19.

Bartels, H. (1993). *Gehölzkunde – Einführung in die Dendrologie*. Verlag E. Ulmer, Stuttgart.

Blanchard, R. (2012). *Field and Laboratory Guide to Tree Pathology*. Academic Press, New York.

Denne, M.P. (1974). Effects of light intensity on tracheid dimensions in *Picea sitchensis*. *Annals of Botany* **38**, 337–345.

Dujesiefken, D. (2001). Die Einsatzgebiete des Zuwachsbohrers und die Wundreaktion im Baum nach Bohrung. In: Dujesiefken, D., Kockenbeck, P. (eds). *Jahrbuch der Baumpflege 2001*, pp. 186–190. Thalacker Media, Braunschweig.

Dujesiefken, D. and Liese, W. (2006). Die Wundreaktion von Bäumen – CODIT heute. In: Dujesiefken, D., Kockerbeck, P. (eds). *Jahrbuch der Baumpflege 2006*, 21–40. Thalacker Media, Braunschweig.

Dujesiefken, D., Peylo, A. and Liese, W. (1991). Einfluss der Verletzungszeit auf die Wundreaktion verschiedener Laubbäume und der Fichte. *Forstwissenschaftliches Centralblatt* **110**, 371–380.

Esau, K. (1977). *Anatomy of Seed Plants*. John Wiley, New York.

Eschrich, W. (1995). *Funktionelle Pflanzenanatomie*. Springer-Verlag, Berlin.

Evert, R.F. (1987). The vascular cambium and secondary vascular differentiation. In: Newman, D.W. and Wilson, K.G. (eds). *Models in Plant Physiology and Biochemistry II*, pp. 77–80. CRC Press, Florida.

Evert, R.F. (2006). Esau's Plant anatomy: Meristem, Cells, and Tissues of the Plant Body: Their Structure, Function, and Development. 3rd edition. John Wiley and Sons, New York.

Evert, R.F. and Eichhorn, S.E. (2013). *Raven – Biology of plants*. 8th edition. W.H. Freeman Company Publishers, New York.

Firestone, J. (2003). *Dendrochronology – Volcanoes of the Eastern Sierra Nevada*. http://www.indiana.edu/~sierra/papers/2003/Firestone.pdf [accessed March 2015].

Jasulka, P. and Blatt, A. (2001). Vergleich der Wundreaktionen an Bohrlöchern vom Resistographen, Teredo-Prüfnadelmesssystem und Zuwachsbohrer. In: Dujesiefken, Kockerbeck, P. (eds). *Jahrbuch der Baumpflege 2001*, pp. 191–194. Thalacker Media, Braunschweig.

Kersten, W. (2001). Infektionsgefahr durch den Einsatz invasiver Diagnosegeräte am lebenden Baum. In: Dujesiefken, D., Kockerbeck, P. (eds). *Jahrbuch der Baumpflege 2001*, 195–202. Thalacker Media, Braunschweig.

Kowohl, T., Kehr, R., Wohlers, A. and Dujesiefken, D. (2001). Wundreaktionen und Pilzbefall im Holzkörper nach Resistograph- und Zuwachsbohrer-Einsatz zur Baumuntersuchung im Bereich von Fäule. In: Dujesiefken, D., Kockerbeck, P. (eds). *Jahrbuch der Baumpflege 2001*, 203–211. Thalacker Media, Braunschweig.

Krabel, D. (1998). Mikroanalytische Untersuchungen zur Physiologie des Baumkambiums von *Thuja occidentalis* L. und *Fagus sylvatica* L. *Forstwissenschaftliche Beiträge Tharandt/Contributions to Forest Sciences* **2**, 1–97.

Ladefoged, K. (1952). The periodicity of wood formation. *Kongelige Danske Videnskabernes Selskab. Biologiske Skrifter* **7**, 1–98.

Langenfeld-Heyser, R. (1989). CO_2 fixation in stem slices of Picea abies (L.) Karst.: microautoradiographic studies. *Trees* **3**, 24–32.

Mellerowicz, E.J., Coleman, W.K., Riding, R.T. and Little, C.H.A. (1992). Periodicity of cambial activity in Abies balsamea. I. Effects of temperature and photoperiod on cambial dormancy and frost hardiness. *Physiologia Plantarum* **85**, 515–525.

Nunez, L.M. (2010). Seasonal dynamics of total soluble Proteins in adult trees of Quercus petraea (Matts.) Liebl. and Fagus sylvatica. *Revista Mexicana de Ciencias Forestales* **1**(1), 75–82.

Oven, P., Schmitt, U. and Stobbe, H. (2000). Wundreaktionen im Splintholz der Gemeinen Kiefer. In: Dujesiefken, D., Kockerbeck, P. (eds). *Jahrbuch der Baumpflege 2000*, 208–211. Thalacker Media, Braunschweig.

Parthasarathy, M.V. and Klotz, L.H. (1976). Palm "Wood" anatomical aspects. *Wood Science and Technology* **10**, 215–229.

Roloff, A. (2013). *Baumpflege*. Eugen Ulmer KG, Stuttgart.

Sachs, T. (1981). The control of the patterned differentiation of vascular tissues. In: Woolhouse, H.W. (ed). *Advances in Botanical Research*, Vol. **9**. Academic Press, New York.

Schwarze, F.W.M.R. (2001). Die Fäuledynamic im Splintholz lebender Bäume. In: Dujesiefken, D., Kockerbeck, P. (eds). *Jahrbuch der Baumpflege 2001*, 171–185. Thalacker Medien, Braunschweig.

Shigo, A.L. (1979). Tree decay an expanded concept. *USDA Forest Service Agricultural Information Bulletin* **419**, 73.

Shigo, A.L. and Marx, H.G. (1977). Compartimentalization of decay in trees (CODIT). *USDA Forest Service Agricultural Information Bulletin* **405**.

Stobbe, H., Schmitt, U., Eckstein, D. and Dujesiefken, D. (2002). Developmental stages and fine structure of surface callus formed after deberking of living lime trees (Tilia sp.) *Annals of Botany* **89**, 773–782.

Thoen, D. and Bâ, A. (1989). Ectomycorrhizas and putative ectomycorrhizal fungi of *Afzelia africana Sm.* and *Uapaca guineensis Mull. Arg.* in southern Senegal. *New Phytologist* **113**, 549–559.

Wareing, P.F. (1956). Photoperiodism in woody plants. *Annual Review of Plant Physiology* **7**, 191–214.

CHAPTER 4

Urban tree roots: Problems and peculiarities

Sandra Korn

Technische Universität Dresden, Tharandt, Germany

4.1 Damages to and influences on the root system of urban trees

While everybody is aware of the aboveground benefits and risks of trees growing in urban environments, the belowground part of these trees is often not taken into consideration. Widely known, of course, are the basic functions of the root system in supporting the tree by uptake of water and nutrients and their transport to the stem. Anchorage of the whole tree is also a very important function of the tree's root system (see Chapter 3.1.2). More than that, however, the root system is "the counterpart of a tree's crown subsurface" (Bernatzky, 1994) and an essential part especially in a challenging surrounding for a tree like urban areas with their many stressors and growth restraints.

4.1.1 Site conditions

Certainly, the root system of urban trees, just like that of trees in their natural habitats, is subjected to numerous influencing site factors. These include soil temperature, moisture and aeration. The optimum temperature for root growth of the most temperate species is around 20°C, due to physiological processes (enzyme activity, metabolism, etc.) (Balder, 1998). However, as it is dependent on the local temperature regime (Day *et al.*, 2010a), it shows a wide variation.

Indirectly, a vigorous and growing root system also depends on soil biological activity (soil fauna, mycorrhiza, diffusion rates, etc.) but, especially in cities, elevated root zone temperatures are often to be found, and temperatures of more than 30°C are frequently observed. Some studies indicate premature senescence and high mortality as a result (Graves and Dana, 1987; Graves *et al.*, 1991).

Comparisons between tree seedlings growing in soils with an elevated temperature (34°C as opposed to 24°C) show different reactions. While (for example) *Gleditsia triacanthos* benefits from elevated soil temperatures, temperate tree species such as *Acer rubrum* or *Ailanthus altissima* are negatively affected. They reduce water uptake rates as well as growth (Graves *et al.*, 1991).

Most tropical trees show less to no reactions to elevated soil temperature, and their growths rates are influenced more strongly by soil moisture (Day *et al.*, 2010a). The root

Urban Tree Management: For the Sustainable Development of Green Cities, First Edition. Edited by Andreas Roloff.
© 2016 John Wiley & Sons, Ltd. Published 2016 by John Wiley & Sons, Ltd.

Table 4.1 Comparison of the composition of soil air at one meter depth (Roberts *et al.*, 2006).

	Paved roadway	Adjacent undisturbed forest	Atmospheric
O_2	4%	18%	21%
CO_2	15%	2%	0.03%

growth of palms, on the other hand, is linearly correlated with soil temperature (Broschat, 1998). Reactions to changes in the soil temperature, and the mechanisms of dealing with these changes, are highly variable.

Looking at the opposite situation, very low temperatures or frost events evoke quite different effects on a tree's root system, mostly affecting exposed newly planted trees and unprotected excavated roots in trenches and at construction sites. Damage to roots may occur at temperatures less than −1°C to −3°C (Balder, 1998). Freezing temperatures cause the formation of ice in the intercellular spaces and in water-conducting tissue (xylem). This leads to damage of cell organelles, disintegration of plasmalemma through death of the cells, up to mechanical rupture of tissues caused by ice crystals. Apparent symptoms are browning of the roots, showing a root necrosis (often damaged cambium and phloem), proceeding at the root collar and the trunk. Consequences for the whole tree range from, at least, delayed leaf formation, microphylly and discoloration of leaves, up to more complex diseases due to limitations in water, nutrient and assimilate supply (Balder, 1998).

Soil moisture and soil aeration are closely linked factors. Both moisture and aeration of the soil depend on the grain and pore size distribution. Macropores (10–50 µm in diameter) are air-filled, while smaller pores retain water. Less than 10% oxygen in soil inhibits the respiration and, therefore, root growth in apple trees (Kozlowski, 1985). Oxygen as low as 7% is tolerable, but less than 5% O_2 in the soil leads to a die-back of roots, microorganisms and fungi. Root growth is restricted by limited aeration because growing root tips have especially high O_2 requirements (Kozlowski, 1985).

How much water is available for plants depends on the texture, organic matter content and structure of the soil. The finer the texture, the less organic matter there is, and the smaller the pores, the more strongly water is bound to particles and is thus not available. Water availability is one of the crucial factors for root growth in urban environments. Specific climatic conditions increase the transpiration of trees and create an elevated water demand. The limited soil volume of the tree pits entails a limited water reserve. Sealing and compaction of soil aggravates this by inhibiting infiltration of precipitation and by changes in soil structure.

Table 4.1 shows a comparison of aeration conditions of an undisturbed forest and a nearby paved roadway (Roberts *et al.*, 2006). Aeration beneath the paved road is severely restricted. A negative impact on root function can be suggested (Morgenroth and Buchan, 2009). The tolerance to poor soil aeration varies among tree species; riparian species like *Populus* spp., *Salix* spp., and *Acer negundo* tolerate poor soil aeration (Kozlowski, 1985).

4.1.2 Human activity

There are several factors, directly linked with any human activity occurring in cities, that influence the development and vitality of roots: compaction, organic pollutants, heavy metals, engine oil, and other contaminants that change soil chemistry. As an example, the pollutant salt is discussed here. At least in northern countries, one of the most detrimental pollutants is *de-icing salt* (see Chapter 10.3.9). In Germany, 1.5 million tonnes yr^{-1} (Raasch and Spengler, 2006) of sodium chloride (NaCl) are spread out during winter on the streets and sidewalks as remedy against glaze and ice. Contamination arises from meltwater, spray and saltwater intrusion into groundwater.

Salt contamination may also be caused by sea salt in coastal regions, with the same effects (Day *et al.*, 2010a): changes in nutrient composition (replacement of potassium and magnesium, two essential nutrients, by Na^+ and Cl^- ions from the salt), and change of the osmotic potential of the rhizosphere. This affects the water and nutrient uptake of trees, even during the growing season. Water is bound more strongly and is therefore less available, leading to drought stress.

An elevated NaCl uptake can be held responsible for a dieback of the root system, especially fine roots, due to a lethal accumulation of Cl^-. More pronounced, however, is damage due to accumulation of salt in the leaves, twigs, winter buds and so on (Day *et al.*, 2010a), leading to necrosis and chlorosis in the leaves, microphylly, early leaf senescence and/or early leaf fall (Balder, 1998).

Physiological mechanisms for tolerance of salt are often analogous or coupled with drought, water logging and cold tolerance (Sæbø *et al.*, 2003), but are highly variable. For instance, *Quercus robur* and *Robinia pseudoacacia* seem to be able to exclude NaCl from uptake. Other species, such as *Ulmus glabra, Salix alba, Gleditsia triacanthos, Elaeagnus angustifolia* and *Picea pungens*, are also known to be salt-tolerant, but the mechanisms are not yet identified (Dobson, 1991; cited from Roberts *et al.*, 2006).

The most important aspect for tree roots in urban environments is *soil compaction* (see Chapter 10.3.8). Soil compaction alters or destroys the soil structure, diminishes porosity and increases strength, resulting in a loss of macro- and medium-sized pores (Kozlowski, 1985; Day *et al.*, 2010a). As discussed above (section 4.1.1), these pores are needed for gas exchange within the soil and are therefore responsible for the supply of O_2 to roots and soil fauna. Consequently, there is a decrease in the root growth and of mycorrhizal activity, due to lack of oxygen. Limited tree growth and vitality was found by Rahman *et al.* (2014), studying *Pyrus calleryana* under such urban soil conditions, and was also reported by Jim (1997) in a large-scale study of tree pit soils in Hong Kong.

Increase in strength (physical resistance to penetration) has a more direct impact on the root system architecture. Roots are shallower and highly branched, with thicker, stubbier roots (Day *et al.*, 2010a), because root growth is prevented or restricted by the soil resistance (Roberts *et al.*, 2006).

Soil compaction can be evoked both intentionally and unintentionally. Soils are compacted to provide structural support during the building of pavements, road beds and foundations (Roberts *et al.*, 2006). The soil of tree-covered parks and playgrounds, soil around shade trees on campsites, parks, and in tree pits normally, is unintentionally compacted as a result of the planned (recreational) use (Kozlowski, 1985).

A study on a visible loss of vitality of mature *Fagus sylvatica* trees on a newly built playground showed a high concentration of CO_2 – conversely leading to a low O_2 concentration – in "tread-on" areas like the sand below children's swings, combined with a reduction in fine roots (Gaertig *et al.*, 2010). Fewer roots induce a deficiency in water and nutrient supply of the crown. Thus, the dieback of branches can be interpreted as the re-establishment of an adjusted root/shoot ratio.

Building of new, heavily trafficked (pleasure) grounds with established old-growth trees should at least be viewed critically, due to the expected soil compaction changing the soil conditions. Trampling damage can, for example, be minimized by applying bark mulch. The vibration of traffic along alleys or streets exacerbates the problem of soil compaction by separation of the grain size (Craul, 1992; cited from Roberts *et al.*, 2006).

Figure 4.1 Highly compacted soil of a tree pit at a densely frequented sidewalk.

Unwanted soil compaction in tree pits is mostly linked to poor tree protection measures (Balder, 1998), caused by entering, parking on and/or driving over the rooting area (Figure 4.1).

There is wide variation in the ability of different tree species to cope with soil compaction. Day *et al.* (2010a) formulated a "root growth opportunity" hypothesis to explain the tolerance of some species for growing in highly compacted soils; soil strength is linked with soil moisture and soil aeration. Species that are tolerant of wet (hypoxic) soils, such as riparian species, may be able to grow roots in urban compacted soil.

The implications of compaction for soil and for trees are nicely summed up by Perry (1978): "No oxygen – no roots; no water – no roots; rigid, impenetrable soils – no roots".

4.1.3 Construction sites

Two main problems are to be found at construction sites. One is the physical destruction of fine and coarse roots of trees in close proximity to a construction site by capping, ripping or pinching (Figure 4.2). Additional damage of the exposed roots may occur due to heat in the summer, or freezing during frost periods. Infection with fungi and spread over the whole tree will be one more consequence.

Other problems affecting the root system near of construction sites are the soil compaction provoked by vehicle movement, stocking of materials (Figure 4.3) and changes in soil level by in-filling. The weight of the fill over the root zone of an existing tree compresses the soil pores, and may act as a barrier to gas exchange between soil and atmosphere, resulting in poor aeration (MacDonald, 2004). The feared consequences of root dieback were rebutted in studies of Day *et al.* (2001) and MacDonald (2004), but they

Figure 4.2 Ripped roots in a trench at a construction site.

Figure 4.3 Stocking of materials during construction work at Großer Garten, a public park in Dresden (left), and at amenity street trees in Stockholm (right).

Table 4.2 Causes of death and damage at urban construction sites (based on a leaflet provided by Reigate and Banstead Council and Lambeth Borough Council (Adapted from Roberts *et al.* 2006).

Cause	Result	Preventive action
Stockpiling of materials or filling soil.	Damage bark, allows infection with pathogens, kill trees.	Most effective: erection of fencing around the tree, including the rooting zone, before any work begins.
Parking or operating machinery under trees.	Damage bark, allows infection with pathogens, kill trees; soil compaction.	Keeping materials, building site huts, vehicles and machinery working or parked away from rooting area (at least tree spun); protect trunks – for example, by wrapping paling to tree.
Traffic over root system.	Soil compaction, anaerobic conditions in rooting area.	
Fires below branches or near trees.	Damage bark, kill tree.	No lighting of fires.
Excavation works, raising or lowering soil level within tree span.	Ripping, capping of roots, diminish stability of tree, allows infection with pathogens.	Excavation near the tree trunk should be done by hand (see section on cable trenching).
Careless lopping of roots or branches.	Stunt or maim the tree, diminishes stability, allows infections with pathogens.	

showed upward growth of new roots in the fill, due to altered water distribution. This may induce tree instability, if older, deeper roots die. Therefore, the erection of root protection zones around trees near construction sites that should be preserved is recommended.

As a rough rule of thumb, the zone should measure a ground radius of 0.18 per cm trunk diameter (Harris *et al.*, 1999; cited from Roberts *et al.*, 2006), but a variety of guidelines are available in different countries (Roberts *et al.*, 2006). Table 4.2 sums up the most important causes of death and damage that one may encounter at construction sites.

Special attention must be given to dealing with cable trenching in the vicinity of street trees. Only a few official guidelines recommend a precautionary area. Roberts *et al.* (2006) state that this should be four times the trunk circumference where precautions to protect the roots are essential. This means that, in this area, there should be only manual excavations, to retain as many roots as possible, although trenchless techniques are to be preferred. Roots should be pruned with a clean, sharp tool, making a clean cut with a small wound. If the trench has to be left open overnight or for many days, exposed roots should be protected with dry sacking or similar, to prevent freezing or heating damage. Roots should be covered with damp sacking, or sprayed with water to prevent drying out during hot days (Roberts *et al.*, 2006).

4.2 Damage caused by the root system of urban trees

When considering the planting or maintenance of trees in urban environments, one will inevitably encounter conflict between the needs of urban infrastructure and the needs of planting trees. While setting up the infrastructure to build robust streets and pavements, with low maintenance costs, means maximum soil compaction and low water conditions,

Figure 4.4 Cracking of asphalt (left) and lifting of curbstones (right) caused by tree roots.

the opposite is true for greening of the streets. Tree pits need filling and covering with good drainage for water, as well as oxygen, along with loose substrate to ensure vital growth. This leads to competition for underground space, resulting in a high possibility of damage to the infrastructure by the roots.

During road construction, the soil is highly compacted in order to reach the necessary load capacity. Roots avoid this highly compacted soil by growing in the looser bedding layers, following the "line of least resistance". The ingrowth of the finest roots initiates a rapid root growth, heading to water and nutrient sources, followed by radial growth of the roots. These have enough force to lift kerb stones, paving slabs and, sometimes, even porches, and to crack asphalt covering (Randrup *et al.*, 2001) (Figure 4.4).

Furthermore, the ingrowth of roots in sewers, drains and pipes can be very problematic. Tree root intrusion causes 50% or more of sewer blockages (DeSilva *et al.*, 2011; Östberg *et al.*, 2012). Here, too, roots grow in the base materials of the pipes, where the soil is less compacted. Additionally, the root growth towards the pipe can be stimulated by condensation water at the pipe's surface, or the leaking of water through damaged pipes (Roberts *et al.*, 2006). Water flowing through sanitary drains, in particular, can be substantially warmer than the surrounding soil, leading to a mass of fine roots forming around the pipe, due to elongated root growth. However, the root intrusion is only possible through physical damage of the pipe's structure or through leaks in joints. Concrete and clay pipes are more often intruded than are modern PVC pipes.

Once roots have entered the pipe, further growth depends on the type of pipe. As roots require oxygen to survive, they will not grow in pipes that are full of stagnant water. Some species adapted to waterlogged conditions, such as *Salix* spp. or *Taxodium* spp., are able to grow in such pipes, enabled by aerenchyma tissue in the roots, but this is an exception (Roberts *et al.*, 2006). Extensive root growth is possible in pipes with a lower water level (Stützel and Bosseler, 2003).

In general, roots often remain constrained to a flattened section at the point where they have entered the pipe, avoiding any direct contact with the water in waste water sewers. Such sewers and drains often carry chemicals from domestic or industrial sources, which are often phytotoxic to some degree. In clean (rain) water drains, on the other hand, roots will often grow in a ring around the inside, inducing a complete blockage of the drain. In general, all woody species are able to grow roots in pipes; it is rather a matter of soil property, distance between tree and pipe, and pipe material (Östberg *et al.*, 2012).

Mechanical or chemical methods can be used for root removal if root ingrowth blocks pipes. Mechanical removal can be done simply by rodding, using connectable rods and a cutter head. Roots can also be cut using high-pressure water cutters and CCTV inspection. These water cutters are able to cut roots up a diameter of 15 cm, and also cut fibrous material blocking the drain (Roberts *et al.*, 2006). This allows very precise cuts to be made, as well as preventing damage to the pipe with the cutter. An additional benefit of this method is that it flushes away the remains of the roots. The disadvantage of this approach is that it is effective only in the short term, because cutting of the roots promotes massive fibrous root development and, consequently, repeated blocking when the defect of the pipe is not repaired concurrently (Rolf and Stål, 1994).

The application of chemicals like herbicides have been shown to control root growth in sewers and drains, and copper sulphate has also been used, for example in the USA. This results in localized root killing in the first instance but, in the long term, such agents have been shown to damage or kill the whole tree and contaminate the environment and, therefore, they are illegal in many countries (Roberts *et al.*, 2006).

4.3 Precautions/preventing damage

There are several ways to protect urban facilities from damage by expanding tree root systems. Different types of root barriers are available to restrict root growth to a more favorable space: traps, deflectors or inhibitors (Randrup *et al.*, 2001; Morgenroth, 2008).

Traps are made of textile fabric which is permeable by water and can be installed below asphalt. The close-meshed fabric minimizes the growth of large diameter roots and therefore minimizes damages to the asphalt or concrete surface.

Deflectors are typically made of fiberglass or plastics that block the roots from lateral expansion. Circular deflectors may be installed around the root ball when trees are newly planted. This kind of barrier allows only vertical root growth. It is suitable for street tree pits and trees growing in restricted spaces. There is some risk that circular barriers may affect and lower the anchorage of the growing tree (Randrup *et al.*, 2001). Linear deflectors, on the other hand, are more favorable. Installation is possible at any time and they are suitable for lawns, planting strips and comparable sites. This barrier blocks the spreading of the roots only in one direction (e.g., at the border to a sidewalk), while unrestricted root growth is possible in other directions.

Some barriers use chemical inhibitors like a gas killing the root hairs or more common copper-soaked fabric. Copper restricts the activity of meristems and cambia, the division-active plant tissues.

Recently, the use of high-density materials has become widespread. Those materials are highly compacted substrates and, therefore, minimize the in-growth of roots by increased resistance. Root growth is directed into the intended space (e.g., the tree pit). Smiley (2008) recommends, as result of study, a combination of linear root barriers and structural soil with compacted fractions.

Several approaches exist to calculate the dimensions of root protection zones for the actual root system. Day *et al.* (2010b) found a strong ratio between trunk diameter and the root system radius, namely 1 : 38. However, this ratio weakens as the tree matures. Additionally, it should be taken into account that urban root systems are not uniformly spread around the tree trunk.

Root intrusion into pipes, sewers and drains can be avoided by careful construction; all joints should be watertight and fitted correctly. Flexible telescopic joints seal sewers against

Figure 4.5 Examples for avoiding compaction of topsoil for street trees. Left: Pervious pavement at busy park. Middle: Tree grid at busy avenue. Right: Ornamental planting at sidewalk.

in-growth of roots. If possible, the best option is replacement with modern, more robust materials. Techniques exist for repairing damaged pipes *in situ* using trenchless technology. For example, a resin-impregnated fabric or deformed high-intensity polythene (HDPE) can be installed in the damaged pipe, expanding and curing the liner into place. It fits snugly on the pipe's interior surface and provides a water- and root-tight seal.

The most effective way to avoid damage to the roots and to the urban infrastructure is planning ahead. Most severe damages to pavements, asphalt and so on occur within a radius of two meters around the tree trunk. Leaving adequate areas of unpaved ground, and improving the pavement design by bridging or raising the pavement above existing roots, can help to keep repair costs to a minimum. Another possibility is to synchronize tree and road life spans; laying of cable or drains and planting of trees may be done at the same time. The life cycle of modern underground facilities and the lifespan of urban trees are approximately similar, at 80–100 years. Therefore, trees can be replaced at the same time as renovation and replacement of the facilities.

Planning ahead includes enhancing the rootable soil volume. Several recommendations or guidelines give attuned advices for tree pit design and dimensions (e.g., DPR, 2008; FLL, 2010; Sydney, 2011; HKPSG, 2014) If tree roots have enough space where they can grow their roots, they will not grow in unwanted areas. The more rooting space is planned, the less damage will occur. An enhanced rooting zone can be reached by quite simple measures:

- Planting trees in rows: increasing rootable soil volume by planting in a continuous trench, rather than in individual pits.
- Planting trees in close proximity to lawns, parks, etc.: Make provisions for growth of roots out of tree pit into soil beneath the grass.
- Avoiding compaction of topsoil in the tree pit: use tree grilles, mulching, bollards, pricks, greening with ornamental plants, and other barriers that prevent walking and parking (Figure 4.5).

4.4 Conclusions

Trees are important and beneficial structures in urban environments, with an expanding part below ground. The root system is constantly growing, and explores new soil areas. It is subjected to many influences, which differ significantly in urban areas compared with natural habitats. One the one hand, root system growth is influenced by human activities and often limited by conditions in the subterranean architecture of cities. The limited space below ground, on the other hand, is the main cause of many conflicts with urban facilities. Enhancing rootable soil volume or controlling growth conditions may solve several of the issues that commonly occur, but this is not applicable for every urban tree. Thus, several questions concerning management of root growth remain open.

References

Balder, H. (1998). *Die Wurzeln der Stadtbäume*. Parey Verlag, Berlin.

Bernatzky, A. (1994). *Baumkunde und Baumpflege*, 5th edition. Thalacker Verlag, Braunschweig.

Broschat, T.K. (1998). Root and shoot growth patterns in four palm species and their relationships with air and soil temperatures. *HortScience* **33**, 995–998.

Craul, P.J. (1992). *Urban soil in landscape design*. Wiley & Sons, New York.

Day, S.D., Seiler, J.R., Kreh, R. *et al*. (2001). Overlaying compacted or uncompacted construction fill has no negative impact on white oak and sweetgum growth and physiology. *Canadian Journal of Forest Research* **31**, 100–109.

Day, S.D., Wiseman, P., Dickinson, S., *et al*. (2010a). Tree root ecology in the urban environment and implications for a sustainable rhizosphere. *Arboriculture & Urban Forestry* **36**, 193–205.

Day, S.D., Wiseman, P., Dickinson, S., *et al*. (2010b). Contemporary concepts of root system architecture of urban trees. *Arboriculture & Urban Forestry* **36**, 149–159.

DeSilva, D., Marlow, D., Beale, D. *et al*. (2011). Sewer blockage management: Australian perspective. *Journal of Pipeline Systems Engineering and Practice* **2**, 139–145.

Dobson, M.C. (1991). *De-icing salt damage to trees and shrubs*. Forestry Commission Bulletin 101. HMSO, London, UK.

DPR (2008). *Tree Planting Standards*. New York City Department of Parks & Recreation. Download from http://www.nycgovparks.org/permits/trees/standards.pdf [accessed Feb. 2015].

FLL (2010). *Empfehlungen für Baumpflanzungen*. Teil 2: Standortvorbereitung für Neupflanzungen; Pflanzgruben und Wurzelraumerweiterung, Bauweisen und Substrate. Forschungsgesellschaft Landesentwicklung Landschaftsbau (ed). Selbstverlag, Bonn.

Gaertig, T., Gerhardt, D. and Weltecke, K. (2010). Nachweis von Bodenbelüftungsstörungen infolge der Neuanlage eines Spielplatzes. In: Dujesiefken, D. (ed). *Jahrbuch der Baumpflege*, pp. 280–285. Haymarket Media, Braunschweig.

Graves, W.R. and Dana, M.N. (1987). Root-zone temperature monitored at urban sites. *HortScience* **22**, 613–614.

Graves, W.R., Joly, R.J. and Dana, M.N. (1991). Water use and growth of Honey Locust and Tree-of-Heaven at high root-zone temperature. *HortScience* **26**, 1309–1312.

Harris, R.W., Clark, J.R. and Matheny, N.P. (1999). *Arboriculture: Integrated management of landscape trees, shrubs and vines*. 3rd edition. Prentice Hall, Upper Saddle River.

HKPSG (2014). *Hong Kong Planning Standards and Guidelines*. Download from http://www.pland.gov.hk/pland_en/tech_doc/hkpsg/sum/pdf/sum.pdf [accessed Feb. 2015].

Jim, C.Y. (1997). Urban soil characteristics and limitations for landscape planting in Hong Kong. *Landscape and Urban Planning* **40**, 235–149.

Kozlowski, T. (1985). Soil aeration, flooding, and tree growth. *Journal of Arboriculture* **11**, 85–96.

McDonald, J.D., Costello, L.R., Lichter, J.M. *et al*. (2004). Fill soil effects on soil aeration and tree growth. *Journal of Arboriculture* **30**, 19–27.

Morgenroth, J. (2008). A review of root barrier research. *Arboriculture & Urban Forestry* **34**, 84–88.

Morgenroth, J. and Buchan, G.D. (2009). Soil moisture and aeration beneath pervious and impervious pavements. *Arboriculture & Urban Forestry* **35**, 135–141.

Östberg, J., Martinsson, M., Stål, Ö. and Franson, A.-M. (2012). Risk of root intrusion by tree and shrub species into sewer pipes in Swedish urban areas. *Urban Forestry & Urban Greening* **11**, 65–71.

Perry, T.O. (1978). Physiology and genetics of root-soil interactions on adverse sites. In: *Fourth North American Forest Biology Workshop*, pp. 77–97. Gainesville.

Raasch, U. and Spengler, B. (2006). *Tausalze – ein Problem bei naturnaher Regenwasserbewirtschaftung.* Download from http://www.emscher-regen.de/service/newsletter/2006_4_Tausalze.pdf [accessed Feb. 2015].

Rahman, M.A., Armson, D. and Ennos, A.R. (2014). Effect of urbanization and climate change in the rooting zone on the growth and physiology of *Pyrus calleryana*. *Urban Forestry & Urban Greening* **13**, 325–335.

Randrup, T.B., McPherson, E.G., Costello, L.R. (2001). A review of tree root conflicts with sidewalks, curbs, and roads. *Urban Ecosystems* **5**, 209–225.

Roberts, J., Jackson, N. and Smith, M. (2006). *Tree Roots in the built environment.* Stationary Office Books, Norwich.

Rolf, K. and Stål, Ö. (1994). Tree roots in sewer systems in Malmö, Sweden. *Journal of Arboriculture* **20**, 329–335.

Sæbø, A., Benedikz, T. and Randrup, T.B. (2003). Selection of trees for urban forestry in the Nordic countries. *Urban Forestry & Urban Greening* **2**, 101–114.

Smiley, E.T. (2008). Comparison of methods to reduce sidewalk damage from tree roots. *Arboriculture & Urban Forestry* **34**, 179–183.

Stützel, T. and Bosseler, B. (2003). Kanal voll: Wenn Bäume in Rohren Wurzeln schlagen. *Rubin* **1**, 58–63.

Sydney (2011). *City of Sydney Street Tree Master Plan 2011 – Part D: Technical Guidelines.* Download from http://www.cityofsydney.nsw.gov.au/__data/assets/pdf_file/0010/130240/PartD-TechnicalGuidelines.pdf [accessed Feb. 2015].

CHAPTER 5

Drought stress: Adaptation strategies

Sandra Korn

Technische Universität Dresden, Tharandt, Germany

5.1 What is stress? – Stress concepts

Classical stress concepts define "stress" as a significant deviation from optimal conditions for life. It is always induced by too little or too much of the respective factor. Schulze *et al.* (2005) give, as an example, the environmental factor "temperature" for two plant species – *Rhododendron ferrugineum* and *Zea mays* – which differ in their natural habitat. The rhododendron is to be found in central and southern European mountains in cold habitats, while maize is a Central American warmer climate plant. Therefore, they also differ in their perception of temperature stress. While *R. ferrugineum* finds optimum temperature for growth and vitality at 12–25°C, *Z. mays* has an optimum of 22–28°C.

Temperatures above and below these span are perceived as stress, and have a negative impact on the state of active life of the plant. Temperatures of, for example, 30°C imply more pronounced stress to cold-adapted *R. ferrugineum* than to maize. Thus, the optimum and stress span of any environmental factor, as well as the stress response of the plant, are highly variable and will depend on species, adaptation, combination of multiple stresses, etc. Also, the duration of the stress plays an important role on how plants can cope with it. A distinction should be made between temporary stress and permanent stress (Larcher, 1995). Temporary stress generally results in short-term impairment, since the stress response of the plant (its functional state) returns to normal after a short period of recovery. In contrast, permanent stress lowers the functional state of the plant over a long period. Even after cessation of the stressing conditions and a recovery period, the former normal functional state of life could not be sustained.

There is a large number of biotic and abiotic factors that are able to induce stress in plants (see Table 5.1). The interplay of numerous stressors and/or successive stresses may lead to enhanced or even diminished stress or, in the worst case, to a limitation of the plant's life (Larcher, 1995).

5.2 Stress responses

Trees, like plants in general, are sessile and cannot escape stressors by moving. Trees live over many growing seasons and, therefore, have developed mechanisms to respond to environmental stressors. They have generated an array of regulatory mechanisms for maintenance of their vital processes (water uptake, photosynthesis, metabolism, etc.) and their survival. If a plant is unable to cope with a stressor (lethal stress), it is restricted

Urban Tree Management: For the Sustainable Development of Green Cities, First Edition. Edited by Andreas Roloff.
© 2016 John Wiley & Sons, Ltd. Published 2016 by John Wiley & Sons, Ltd.

Table 5.1 Examples for biotic and abiotic stressors (Adapted from Larcher 1995).

Stressors	
Abiotic	**Biotic**
Radiation • Excess, shading, UV radiation	Plants • Crowding, shading, allelopathy, parasites
Temperature • Heat, cold, frost	Microorganisms • Viruses, bacteria, fungi
Water • Drought, flooding, stagnant water	Animals • Grazing, trampling, urine
Gases • Oxygen deficiency, methane-, CO_2-enrichment	Anthropogenic • Pollution, soil compaction, soil sealing, fire, agrochemicals, waste • Damage, pruning/cutting
Minerals • Nutrient deficiency, salinity, acidity, alkalinity, heavy metals	
Mechanical effects • Wind, snow cover, burial	

to areas without this particular stress factor (e.g., trees which cannot cope with water logging situations do not grow in flood plains). In general, adaptation to stress is a relative measure describing the tree's capability to cope with a stressor relative to other species (Niinemets, 2010).

5.2.1 Adaptation to drought stress – stress escape

For the purpose of discussing the different adaptation strategies, the focus here is on drought as a stressor. Drought may appear as short, dry periods, extended dry seasons in monsoonal climates, or even water/irrigation restrictions. It always includes limitation on water availability. However, it should be kept in mind that these stress concepts are also valid for other stressors, such as salinity, cold, pollution, and so on.

Species adapt to stress using different strategies. The strategy of *escaping* is very common among herbaceous plants, but is not to be found among tree species. Plants use phenological plasticity to avoid stressing periods; geophytes, bulb and tuber geophytes and annual plants survive drought periods by production of seeds or perennating (underground) organs like bulbs, rhizomes, or tubers which are protected from desiccation deep in the soil. They sprout or germinate at appropriate seasons, after rainfall events, or during rainy periods (Bresinsky *et al.*, 2008).

A more important mechanism to cope with stress for trees is stress *resistance*. This adaptation to stress has numerous different characteristics at physiological, cellular, genetic, molecular, metabolic, anatomical and/or morphological levels. Based on different reactions, stress resistance is subdivided once more into tolerance and avoidance strategies (Bresinsky *et al.*, 2008):

- Avoidance mechanisms ensure that essential organs, physiological processes and cytoplasmic structures of the cells are not impaired by any water limitations.
- Tolerance strategists, on the other hand, endure drought stress without damage to the aforementioned parts (Lösch, 2001).

5.2.2 Adaptation to drought stress – stress resistance by avoidance

A broad range of structural and physiological adaptations have evolved to avoid mild or severe drought stress. Tree species each possess a characteristic combination of *avoidance mechanisms*. The two main objectives for this strategy are:

1 to improve the uptake of available water; and
2 to reduce the transpirational water loss.

A deep tap root, in combination with a rapidly growing, extensive root system, allows the tree access to water in deeper soil layers. For example, some *Acacia* species in Australia are found to have roots reaching down to 12 m or more (Moore, 2013). Structural adaptations in the hydraulic architecture of the water-conducting system can also enhance water uptake. Producing more xylem with smaller vessels, enlarges the area of the water-conducting system, concurrently lowering the threat of xylem vessel cavitation (disruption of water flow by the formation of vapor cavities due to insufficient water supply).

Trees have evolved a wide variety of modified leaf characteristics, including alterations in morphology, leaf area reduction, and stomatal control, to reduce the transpirational loss of water. A reduction in leaf area can be achieved by developing pinnate or deeply lobed leaves, or needles and needle-like leaves. These leaf shapes reduce the area exposed to the sun, and have lower leaf temperatures than undivided leaves. A dense hairiness and shiny (upper) leaf surface area have the same effect (Figure 5.1).

An early leaf senescence, as well as leaf abscission during drought periods (e.g., dry-deciduous species from tropical dry forests), is another mechanism that allows trees to reduce their transpirational area.

Figure 5.1 Examples for morphological drought adaptations to avoid drought stress. Left: Pinnate and shiny leaves of *Catania* spec. Right: Dense hairiness of young *Rhododendron* spec. leaves.

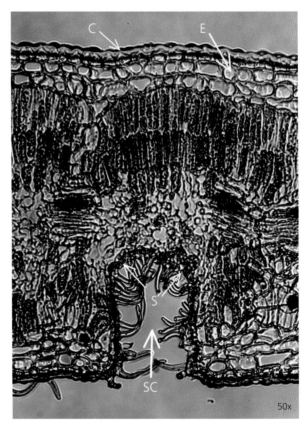

Figure 5.2 Cross-sectional illustration of drought adapted xeromorphic leaf of *Nerium oleander*. C – thick waxy cuticle; E – three-layered epidermis; S – stomata in SC – stomatal crypt with hairs.

Anatomical adaptations at leaf level often exhibit xeromorphic traits (Raven *et al.*, 1988). Figure 5.2 shows a cross-section through a xeromorphic leaf of *Nerium oleander*. These leaves develop a thick cuticle and compacted epidermic cell walls, lowering the transpiration of the epidermic cells. The epidermis can be multi-layered. In addition, the upper leaf surface is covered with thick layers of wax, which give the leaf its character-istic sheen. Stomata are at the underside of the leaf. These are sunken stomata, at the base of pits called stomatal crypts. The stomata are hidden behind miniscule hairs in order to create a moister microclimate inside. Also, the size and distribution of stomata commonly varies with limited soil water availability; the number and size of stomata are inversely related. Plants growing under drier conditions often show smaller, but more densely distributed, stomata. This allows plants to maintain high rates of gas exchange when all stomata are open, but a more pronounced control of stomatal aperture during water-demanding periods. Moore (2013) reports a range from 28 stomata mm^{-2} in arid *Persoonia* species to 100–350 mm^{-2} in *Eucalyptus* species in Australia.

A very important physiological mechanism in avoiding drought stress for trees is the stomatal regulation of the transpiration. Transpiration is the loss of water from stems and leaves, the major part of which is transpiration through stomata. Plants open the stomata to take up carbon dioxide for photosynthesis. At the same time, water vapor diffuses out of the leaf, driven by a gradient in humidity between the saturated atmosphere inside the leaf and the less humid surrounding air. Due to the cohesive character of the water molecules, this process moves water as a continuous column through the whole tree. The driving force is a water potential gradient, where "water potential" describes the energetic state of the water retained by physical and chemical forces in cells, organs, soil, etc. (simplified: their "capability for water uptake"). Water potential is negative by definition, and becomes more negative with lack of water.

Water moves from areas of higher water potential (e.g., saturated atmosphere inside the leaf) to areas with lower water potential (e.g., less humid surrounding air). This is the driving force for water uptake, water movement through plants, and their transpiration, too, and is called SPAC (soil plant atmosphere continuum). By controlling the aperture of the stomata, the tree has the option to control the amount of transpirational water loss. The aperture of the stomata is expressed as stomatal conductance. The higher the stomatal conductance is, the more water vapor evaporates through the stomata (Lösch, 2001).

Trees avoiding drought stress close their stomata partially or completely during periods with limited water availability and/or enlarged atmospheric evaporative demand (high temperature, direct solar radiation, low air humidity, high VPD). Thus, they have control over the water balance. Consequently, stomatal conductance is lowered before any negative impacts on leaves by water shortage occur. Transpirational water loss is prevented, but water potentials are less negative and, thus, water uptake is limited (Lösch, 2001).

The consequences for the tree's water balance demonstrate an *isohydric* response: lowered stomatal conductance, reduced rates of gas exchange O_2/CO_2, less negative water potential of the tree, and a reduction or cessation of water uptake. This slows down the depletion of the available water (Kjelgren *et al.*, 2013), as well as minimizing the loss of water. It also leads to the required consistency of the tree's water balance without large variations in water potential. On the other hand, growth of the tree is also reduced (Cowan, 1997; Moore, 2013).

5.2.3 Adaptation to drought stress – stress resistance by tolerance

Tree species using *tolerance* mechanisms to cope with drought stress are also referred to as *anisohydric*. Tolerators keep their stomata open and are luxury water users during limited water availability (Moore, 2013). They endure water limitations in their tissues that are caused by the high transpirational water loss. Opposed to avoiders is the maintenance of the gas exchange (uptake of CO_2), which is of greater importance than reducing the water loss. Stomata remain open, and consequently there is high stomatal conductance, attended by high transpiration rates. Water potentials in the tree decrease, and the water potential gradient is steepened in order to ensure a high uptake rate of more strongly bound soil water. Trees have to tolerate high variations in potentials, and particularly they have to cope with a possible loss of cell turgor in leaf tissues. Thus, to tolerate drought stress means to handle and survive cytoplasmatic dehydration.

Osmotic adjustment allows tolerant tree species to maintain their cell turgor completely, or at least partially to steepen the water potential gradient (Lösch, 2001). This takes place at the cellular level, with the cell's vacuole as an important part which is filled with water and soluble compounds (among others). The vacuole maintains and regulates the internal hydrostatic pressure – the turgor – of the cell. Without limiting

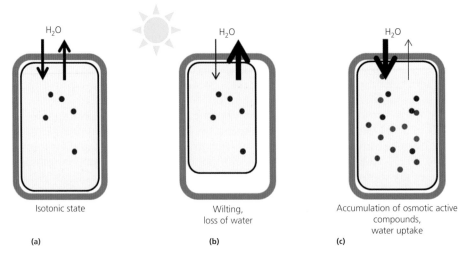

| Isotonic state | Wilting, loss of water | Accumulation of osmotic active compounds, water uptake |
| (a) | (b) | (c) |

Figure 5.3 Simplified presentation of water status and osmotic active compounds during **a**) isotonic state of the cell, **b**) water loss, and **c**) osmotic adjustment.

water conditions, the plant cell is in an isotonic state, with osmotically driven water influx and efflux (Figure 5.3a). Dehydration leads to loss of water in the vacuole, resulting in a wilting process (Figure 5.3b). This is prevented by the accumulation of osmotically active compounds such as amino acids, sugar, sugar alcohols within the vacuole (Figure 5.3c). The osmotically driven inflow of water slows down the loss of turgor and protects the cell membranes and proteins.

Osmotic adjustment is an important process in the plant's resistance to drought stress. It extends the period and improves the turgescence of cells and tissues of the tree. Thus, tolerant trees are able to maintain photosynthesis, transpiration and water uptake during periods of water limitation. As a quite rough generalization, it can be stated that anisohydric/tolerant species, including palm species, seem to be found more frequently in drought-prone habitats (Gomes and Prado, 2007) than do isohydric/avoiding species (McDowell et al., 2008).

Table 5.2 summarizes the discussed parameters of anatomical, morphological and the most important physiological adaptation mechanisms of the avoidance and tolerance strategy.

There are no generalized thresholds or ranges for the two described categories of drought stress resistance mechanisms. Furthermore, it may be observed that a large number of potential interactions and overlap are possible. Many of the responses may be adjusted, either to tolerate or to avoid drought stress within species and/or site-specific limits (McDowell et al., 2008). Both response mechanisms may cause damage or mortality by "hydraulic failure" or by "carbon starvation". Hydraulic failure is more likely to affect anisohydric species during intense droughts. This is caused by an exceeding (evapo)transpiration, with too-negative water potentials and xylem cavitation.

On the other hand, isohydric species are more vulnerable to carbon starvation during prolonged but intermediate drought periods. They close the stomata to prevent water loss, but also restrict carbon gain from photosynthesis. Ongoing metabolism and respiration deplete carbohydrate reserves without the possibility of replenishing, making them more susceptible to insect and pathogen attacks (McDowell et al., 2008).

Table 5.2 Short summary of avoidance and tolerance mechanisms showing an adaptation to drought stress.

Avoidance		Tolerance	
Adaptation	**Response**	**Adaptation**	**Response**
Improvement of water uptake	Extensive, deep root system	High tolerance against plasmatic dehydration	
	Increase of conducting system	Osmotic adjustment	Maintaining cell turgor
Reduction of water loss	Anatomical and morphological adaptations		
	Stomatal regulation (closing)		
	Leaf abscission, senescence		
Physiological consequences	Isohydric: • Reduction of transpiration and photosynthesis • Low stomatal conductance • Less negative water potential • Stop water uptake	Physiological consequences	Anisohydric: • Steepened water potential gradient • More negative water potential • Higher stomatal conductance • Maintaining transpiration and photosynthesis • Maintaining water uptake

5.3 Identifying tree species adapted to stress

5.3.1 Responses and adaptations to drought stress

Responses and adaptation parameters to drought stress, as described in the previous section, can be subdivided into short-term/temporary, intermediate, and long-term responses to drought stress (Roloff, 2013). This classification depends on the duration of the drought stress exposure, the duration of the response time, when the response is called for, and the provenance of the tree. All responses and adaptations, of course, are part of the genetic constitution of the tree, and are activated when needed.

Short-term responses are a direct response to current drought events with a (relatively) short duration. These responses usually cease with the end of the drought (onset of precipitation, start of wet season in monsoonal climates, irrigation of urban sites, etc.). Intermediate responses can be observed during and/or after drought periods, and are often an adaptation to frequent or recurring drought events. In contrast are long-term adaptations, which are not triggered by current drought events or periods but are adaptations to dry habitats. These are often typical xerophytic morphological-anatomical characteristics (Table 5.3).

5.3.2 Identifying suitable tree species

Knowledge of stress adaptation and performance of a tree during drought events is – apart from its aesthetic and functional aspects – an important base for selection of suitable urban tree species (Sæbø et al., 2003). For most species planted in urban environments, very little is known of their response to the stressor drought (see Chapter 15), or if they are stress avoiders or tolerators, let alone their stress physiology.

Table 5.3 Parameters illustrating responses and adaptations to drought stress, with examples for species expressing the described parameter (cited from Roloff, 2013; Gomes and Prado, 2007; Günthardt-Goerg and Vollenweider, 2007; Sjöman and Nielsen, 2010; May *et al.*, 2013; Moore, 2013).

Short-term	Intermediate	Long-term
Stomatal closure (*Casuarina littoralis*)	Short shoots within the crown following a drought year (*Fagus sylvatica*)	Leaves: • Pinnate or deeply lobed (*Quercus frainetto*) • Shiny with thick wax layer (*Koelreuteria paniculata*) • With hairs (*Sorbus aria lutescens*) • Small, lanceolate (*Olea europaea*) • Needle-like (*Casuarina littoralis*) • Visible and palpable veins (*Quercus ilex*) • Sunken stomata (*Banksia* spec.)
Decrease of water potential (*Quercus rubra*)	Dieback of twigs (*Platanus* x *acerifolia*)	Shoot/stem: • Stipules or thorns (*Robinia pseudoacacia*) • Photosynthetic active (*Ulex europaeus*) • Thick epidermis or bark (*Phellodendron amurense*) • Stem water reserve (*Sabal palmetto*) • Adaptation of wood anatomy
Increase of hormone concentration (*Acer platanoides*)	Dieback of fine and small roots (abandoning of branching to retain and supply remaining parts) (*Pinus pinea*)	Root system: • Deep-reaching tap root (*Eucalyptus camaldurensis*) • Two-layered root system (*Pinus* spp.) • Low shoot/root ratio (*Eucalyptus camaldurensis*) • Clonal growth by ground layering or root suckers (*Ailanthus altissima*)
Osmotic adjustment (*Acacia aneura*)	Enlarging stomatal density (*Quercus robur*)	
Leaf rolling, folding, movement (*Fagus sylvatica, Acacia mearnsii*)	Browning and shedding of needles in the months after drought (*Abies* spec.)	
Heliotropism (*Robinia pseudoacacia*)		
Leaf and twig abscission/ facultative and dry-deciduous (*Ficus superba*)		
Leaf senescence (*Acer saccharinum*)		
Decrease of stem diameter (use of stem water reserves) (*Picea abies*)		
Cavitation of xylem vessels (*Acer negundo*)		

There are different approaches to bypass this deficit in knowledge. Kjelgren *et al.* (2013) developed a conceptual framework to show tree adaptations to drought in tropical and subtropical cities. They differentiated between wet evergreen species growing in aseasonal wet climates, dry deciduous and dry evergreen species originating from monsoonal climates. They also highlighted physiological differences in drought adaptation as a crucial factor for choosing and managing tropical and subtropical species in urban environments.

Different from temperate regions, the climate of most subtropical and tropical cities is characterized by one or more dry periods, evoking drought stress. Evergreen species from monsoonal dry forests are adapted to intense solar radiation, heat and drought during dry season, by having small leaves and a deep root system, tolerating low leaf water potentials, low stomatal conductance and lowered photosynthetic rates. Evergreen species from equatorial wet forests, in comparison, are less drought tolerant. Therefore, it is quite reasonable and common that several urban tree species in cities in equatorial wet climates, such as Singapore, originate from monsoonal forests.

On the other hand, some surveys of urban trees in Bangkok (Thailand) and Bangalore (India) showed that dry deciduous species are more adapted to monsoonal dry urban climates than evergreen species (Kjelgren *et al.*, 2013). However, against the background of physiological studies of three common urban tree species in Bangkok (dry deciduous *Lagerstroemia loudonii*, dry evergreen *Swietenia macrophylla* (mahogany), evergreen/facultative dry-deciduous *Pterocarpus indicus*), it was concluded that monsoonal evergreen species are also adapted to the subtropical and tropical urban environment. The physiology of the three species revealed that they follow an avoidance strategy similar to temperate avoiding tree species. Australian tree species, such as some *Acacia* and *Eucalyptus* species, or the evergreen *Casuarani littoralis*, are also identified as avoiders.

Palms follow the same mechanisms as other woody tree species (Schmitt *et al.*, 1993; Pittenger *et al.*, 2009). Commercially used palm species, such as *Cocos nucifera* varieties, have been characterized as drought-tolerant in their physiology (Gomes and Prado, 2007), but less is known about palm species in arid and urban landscapes. For example, two species (*Washingtonia filifera, Erythea armata*) in a Mexican desert showed a delayed onset of drought stress-applying adaptation mechanisms, such as osmotic adjustment, reflective waxy layers on leaves and a high transpiration rate (Schmitt *et al.*, 1993).

For palms with big, unlobed leaves, high transpiration rates are of particular importance for leaf cooling. Pittenger *et al.* (2009) studied five ornamental, urban landscape palms, including *Washingtonia filifera, Trachycarpus fortunei* and *Syagrus romanzoffiana*, for vitality aspects during drought. None of the palm species showed any serious reductions in growth or aesthetic quality. Of particular note was the statement that these landscape palms need less water than commercially used palms.

As Moore (2013) pointed out, adaptation to drought is often accompanied by a growth reduction during periods with low water availability. Even if urban trees are not grown for timber production, this may affect their usability for urban greening. Benefits of urban trees, such as carbon sequestration and transpirational cooling may be limited, while reduction of leaf area may change the desired canopy characteristics and ornamental function. Therefore, there is a great need to gather information on long-term responses of trees planted in urban environments to drought.

This data is limited to commonly used tree species. However, much more popular and easier to access is the use of knowledge and experience from long-term existing plantings in cities (Sjöman and Nielsen, 2010). This can be called "an informal selection process" (Kjelgren *et al.*, 2013). Street tree lists, like the ones issued by UTF (Urban Tree Foundation www.urbantree.org) or GALK (Gartenamtsleiterkonferenz www.GALK.de), provide this information for practitioners and arborists. The parameters shown in Table 5.3 may be used to identify drought resistant tree species that may be suitable for planting in stressful urban environments. If based on easily accessible parameters, such as anatomical-morphological characteristics, an evaluation may show valuable indicators for drought adaptation (Roloff, 2013). However, it should be kept in mind that these are mostly avoiding, rather than tolerating species, and information on physiological backgrounds is lacking.

5.4 Conclusions

Scarcity of available soil water and a high atmospheric water demand (VPD) contribute to drought stress for plants. It is an important factor for growth and vitality of urban trees. Responses to drought, and mechanisms of dealing with the stressor, are highly diverse and are determined by species, duration and severity of drought events, natural habitat, and others. The responses can be found at biochemical, physiological, anatomical and/or morphological level. A distinction can be made between trees that are avoiding or tolerating drought stress.

Species avoid drought damages by improvement of water uptake and reduction of water loss. They maintain an isohydric water status regarding water potential. Species with a high tolerance against dehydration exhibit an anisohydric water status, with negative water potentials for maintaining water uptake. Both strategies are adaptations to drought stress, and are to be found amongst urban tree species. While avoiding species often have a reduced growth and vitality during prolonged droughts, tolerant species are more affected by intense drought events. Therefore, particular attention should be paid to this issue during the selection of urban tree species. The climate of the urban environment, with respect to the climate zone of the city, is an important factor determining the drought event frequency and severity. Surveillance of long-term planted species, as well as paying attention to morphological/physiological adaptation parameters, provide good indications for drought-adapted tree species.

References

Bresinsky, A., Körner, C., Kadereit, J.W., *et al.* (2008). *Strasburger – Lehrbuch der Botanik*, 36th edition. Spektrum Verlag, Heidelberg.

Cowan, I.R. (1997). Stomatal behavior and environment. *Advances in Botanical Research* **4**, 117–228.

Gomes, F.P. and Prado, C. (2007). Ecophysiology of coconut palm under water stress. *Brazilian Journal of Plant Physiology* **19**, 377–391.

Günthardt-Goerg, M.S. and Vollenweider, P. (2007). Linking stress with macroscopic and microscopic leaf response in trees: new diagnostic perspectives. *Environmental Pollution* **147**, 467–488.

Kjelgren, R., Joyce, D. and Doley, D. (2013). Subtropical-tropical urban tree water relations and drought stress response strategies. *Arboriculture & Urban Forestry* **39**, 125–131.

Larcher, W. (1995). *Physiological Plant Ecology*, 3rd edition. Springer, Berlin.

Lösch, R. (2001). *Wasserhaushalt der Pflanzen*. Quelle & Meyer Verlag, Wiebelsheim.

May, P.B., Livesley, S.J. and Shears, I. (2013). Managing and monitoring tree health and soil water status during extreme drought in Melbourne, Victoria. *Arboriculture & Urban Forestry* **39**, 136–145.

McDowell, N., Pockman, W.T., Allen, C.D. *et al.* (2008). Mechanisms of plant survival and mortality during drought: why do some plants survive while others succumb to drought? *New Phytologist* **178**, 719–739.

Moore, G.M. (2013). Adaptations of Australian tree species relevant to water scarcity in the urban forest. *Arboriculture & Urban Forestry* **39**, 109–115.

Niinemets, Ü. (2010). Responses of forest trees to single and multiple environmental stresses from seedlings to mature plants: Past stress history, stress interactions, tolerance and acclimation. *Forest Ecology and Management* **260**, 1623–1639.

Pittenger, D.R., Downer, A.J., Hodel, D.R. *et al.* (2009). Estimating water needs of landscape palms in Mediterranean climates. *HortTechnology* **19**, 700–704.

Raven, P.H., Evert, R.F. and Curtis, H. (1988). *Biologie der Pflanzen (Biology of Plants)*. de Gruyter, Berlin.

Roloff, A. (2013). *Bäume in der Stadt*. Eugen Ulmer KG, Stuttgart.

Sæbø, A., Benedikz, T. and Randrup, T. (2003). Selection of trees for urban forestry in the Nordic countries. *Urban Forestry & Urban Greening* **2**, 101–114.

Schmitt, A.K., Martin, C.E., Loeschen, V.S. *et al.* (1993). Mid-summer gas exchange and water relations of seven C_3 species in a desert wash in Baja California, Mexico. *Journal of Arid Environment* **24**, 155–164.

Schulze, E.-D., Beck, E. and Müller-Hohenstein, K. (2005). *Plant Ecology*. Springer, Berlin.

Sjöman, H. and Nielsen, A.B. (2010). Selecting trees for urban paved sites in Scandinavia – A review of information on stress tolerance and its relation to the requirements of tree planters. *Urban Forestry & Urban Greening* **9**, 281–293.

CHAPTER 6

Aspects of urban tree pathology

Rolf Kehr

HAWK Hochschule für Angewandte Wissenschaft und Kunst, Fakultät Ressourcenmanagement, Göttingen, Germany

6.1 Definitions, terms and concepts

Some texts that deal with trees suitable for urban environments describe particularly hardy species as being "resistant to pests and pathogens". However, every tree genus and species is subject to diseases and pests in its natural habitat. Compared to forest ecosystems, the focus of tree health problems in urban situations is different, due to differences in climate and in the biodiversity of antagonists, natural enemies, pests and diseases. Also, the same diseases and disorders may be viewed differently in a city than in a forest, because of the requirements posed by public safety or aesthetic demands. One decisive aspect for the management of urban trees is if a disease or a disorder is particularly conspicuous or damaging, if it causes depreciation in value, or if it leads to concerns regarding safety or roadworthiness.

The following chapter aims to give an introduction to the types of pests, diseases and disorders which can affect urban trees and woody ornamentals. In this context, it can only present the concepts involved and provide examples for common problems relevant to tree health. Detailed information on diagnosis and damage potential for certain regions and climates around the world must necessarily be taken from the specific literature for that region. Fortunately, there is a wealth of illustrated literature available that enables tree managers to identify some of the more common or distinct problems, or to categorize a disorder before, perhaps, seeking help from a specialist (see recommended literature). Increasingly, the internet and mobile services offer sites and applications that can also help diagnose problems in the field.

In woody plants, the term *disease* typically applies to infectious diseases caused by plant-pathogenic viruses, bacteria and fungi. These are accordingly termed *pathogens*. *Abiotic disorders* are caused either by external, non-living factors, or by abnormal physiological processes in the host plant. *Pest* is the term for organisms belonging to the animal kingdom, and their attack on woody plants is termed an *infestation*. Diseases and infestations can lead to *damage* if they cause noticeable harm to the plant's health, but this term is often also used to denote the economic effects of a disorder, even if the plant itself does not suffer long-term effects. An example would be the horse chestnut leaf miner (*Cameraria ohridella*), which causes conspicuous browning of leaves and, thus, negatively influences our aesthetic concept of how a city tree should look, but generally has only a minor impact on the tree's growth rate and overall health.

Urban Tree Management: For the Sustainable Development of Green Cities, First Edition. Edited by Andreas Roloff.
© 2016 John Wiley & Sons, Ltd. Published 2016 by John Wiley & Sons, Ltd.

When dealing with pests and diseases of urban trees, it is important to know if the causal organism can pose a threat to other tree species in the vicinity. This knowledge provides an essential basis for disease management and replanting. In forest ecosystems, the epidemiology of major pathogens is an important aspect, where factors such as pathogenicity, host tree characteristics and climatic abnormalities interact and can lead to an epidemic. Depending on the biology of the causal organism and the presence of vectors, some diseases can also spread rapidly within tree species in urban environments. One example is Dutch elm disease (DED), which is caused by the fungus *Ophiostoma novo-ulmi* (Tattar, 1978; Manion, 1981; Tainter and Baker, 1996).

Many disease-causing agents of woody plants are of a secondary nature, and may be referred to as *weak pathogens*. Under normal conditions, these are able to cause extensive damage only after the host has been subjected to some sort of predisposing stress – for instance, abiotic factors. When a large number of a certain tree species is affected by several different pests and pathogens, the result is usually termed a *decline* disease. These are multi-factorial diseases, in which a combination of primary and various secondary factors leads to a decline in the tree's health and to higher mortality levels in the tree population. Most known decline diseases affect forest ecosystems – for example, Beech bark disease – but urban tree populations can also be involved if a tree species dominates a given area, or grows in close proximity to affected forests (Manion and Lachance, 1992; Houston, 1994; Butin, 1995).

6.2 Abiotic damage and disorders

Abiotic disorders are responsible for a high proportion of health problems and damage to trees in urban environments, weakening the trees and predisposing them to organisms that take advantage of their reduced vigor. Difficult site conditions often prevail, in addition to the warmer and drier climatic conditions encountered in cities. Depending on local and site conditions, abiotic factors can include intense light, lack of light, hail, chilling and freezing damage, heat injury and sunscald, flooding or drought. Air pollution, salt (de-icing salt, salt spray in coastal areas, salinity of soils), soil aeration deficits, herbicide damage or physiological disorders caused by nutrient deficiency, are further factors (see Chapter 4, section 4.1.2, and Chapter 10, sections 10.3.8/9). Many of the symptoms exhibited by affected plants are non-specific, making diagnosis of the exact cause, and differentiation from biotic factors, difficult (Costello *et al.*, 2003).

Sunscald is a frequent problem in thin-barked species and young trees in the first years after planting, characterized by vertical zones of dead bark and cambium and subsequent production of callus and woundwood (Figure 6.1). Some species and cultivars of maples (*Acer* spp.) and Linden trees (*Tilia* spp.) appear to be especially susceptible. Deep planting, heavy flush-cut pruning and injuries to the trunk and root, as well as neglected watering of young trees, are major causes. With age, sunscald usually becomes less of a problem, due to maturing and thickening bark and to an increase in vigor after the critical planting or transplanting phase. Trees with sunscald often also exhibit secondary problems, such as infestations by bark insects such as Buprestid beetles, or infection by canker- or decay-causing fungi. Throughout many parts of the world, various methods are applied to protect the stem of freshly planted trees from sunscald for a number of years (Harris *et al.*, 2004).

Inadequate site conditions can lead to physiological disorders such as gummosis, characterized by production of an excess of gum from buds and from the bark, which occurs in species of cherry (*Prunus* spp.). In areas where de-icing salt is used, trees often exhibit leaf necrosis symptoms (Figure 6.2), which may be confused with drought

Figure 6.1 Young roadside *Sorbus intermedia* with sunscald damage.

Figure 6.2 Leaf of *Platanus x hispanica* with chlorosis of the leaf margin after uptake of NaCl.

symptoms. Lack of water is one of the most important abiotic factors (see Chapter 3) that encourage disease development and pest attack in urban trees. Many studies have shown the association between drought and disease by secondary or weak pathogens that cause cankers or diebacks (Desprez-Loustau *et al.*, 2006).

Prevention of abiotic disorders starts with planting suitable species at the right site, taking soil characteristics and climatic conditions into account. Some soil-borne diseases, such as *Heterobasidion* root and butt rot and *Phymatotrichopsis omnivora*, are promoted by calcareous soils with elevated pH-levels, or by highly fertile soils. Planting at the correct depth in suitable soil or substrate with known and well-defined chemical and physical properties, and avoiding mechanical injury, as well as watering and fertilizing adequately, are important prerequisites for maintaining a healthy tree. For palms, it is advisable to consult the specific literature regarding abiotic disorders (Chase and Broschat, 1991; Costello *et al.*, 2003; Elliott *et al.*, 2004).

6.3 Virus diseases

Plant viruses are present in all ecosystems worldwide, and affect many different genera of trees used for orchard production, forestry and urban plantings. However, the assessment of potential damage resulting from infection is difficult. Infections are transmitted via seeds, pollen and plant-feeding insects, and also through cultural practices such as pruning and grafting. Virus infections have been reported for at least 17 genera of tree species, some of which, including *Aesculus, Betula, Carpinus, Fagus, Fraxinus, Populus, Prunus, Robinia, Sorbus,* and *Ulmus,* have been widely planted in urban environments. Symptoms often consist of color changes and mottling in leaves and needles. Affected trees sometimes show a loss in vigor that may be enhanced by other abiotic and biotic causes.

Most economic damage occurs in fruit trees, especially in *Citrus* and *Prunus*. Cherry leaf roll virus (CLRV) is latently present in several forest and urban tree species and, for example, leads to dieback and decline symptoms in *Betula* spp. in Scandinavia (Cooper, 1993; Büttner *et al.*, 2013).

Since there are no feasible curative measures for virus-infected trees, the focus in management must be to identify plant genera and species where viruses contribute to decline diseases and to prevent spread, especially through hygienic procedures in nursery production and in tree care. Worldwide, several certification schemes for virus-free planting stock have been implemented.

6.4 Diseases caused by bacteria and other prokaryotes

Several classes and genera of prokaryotes are relevant to urban trees. Phytoplasmas, obligate plant parasites that belong to the eubacteria, are the cause of disorders in such tree genera as *Alnus, Fraxinus* and *Ulmus*. The diseases are often termed "yellows", and the symptoms consist of chlorosis, leaf curl, premature autumn discoloration, and decline symptoms. Lethal yellowing of palms is present in the southern USA, Mexico and Central America. The disease has a broad host range among palms not native to the US and Central America, making it a threat to many ornamental palms in their native habitats in other parts of the world (Lee *et al.*, 2000; Harrison and Jones, 2004; Griffiths, 2013).

Bacteria are responsible for many known plant diseases, and various tissues and organs may be affected. Symptoms can range from bacterial leaf scorch, shoot diseases and stem cankers, to witches' brooms, galls and tumors. Crown gall (*Agrobacterium tumefaciens*) affects many plant species and causes galls of different appearances in roots and the root collar region, but also on branches and trunks. The bacterium is present in soil and is transmitted by cultural procedures such as propagation in nurseries and pruning. Bacteria, some of them anaerobic, are also present in wetwood in several tree species (Schink *et al.*, 1981; Griffiths, 2013).

Fire blight, caused by *Erwinia amylovorum*, is an invasive bacterial disease with worldwide significance. It affects only members of the plant family *Rosaceae*, some of which, like *Malus* and *Pyrus*, are highly susceptible (Figure 6.3). Originating in North America, the disease spread to New Zealand in 1919 and to Europe in 1956/57, causing heavy economic damage to fruit tree production. Fire blight has high relevance for woody plants in urban environments, since many ornamental varieties of fruit trees belong to the *Rosacea* (Vanneste, 2000).

A serious bacterial disease of horse chestnut (*Aesculus* spp.) has developed in recent years in Europe. First reports associated the bleeding cankers with *Phytophthora* disease but, eventually, a bacterium was found to be the cause of the disease, now called Pseudomonas bleeding canker. *Pseudomonas syringae* pv. *aesculi*, a strain *of P. syringae* specific to *Aesculus* species, apparently developed from a pathogen causing leaf spots on *A. indica* into one causing bark and cambium damage in other *Aesculus* species, notably many of those used in urban plantings throughout Europe. The infection also reduces

Figure 6.3 Fire blight (*Erwinia amylovorum*) led to the death of this *Mespilus germanica*.

host resistance and leads to secondary infection by wood-decay fungi and, hence, the felling of many older trees for safety reasons (Webber *et al.*, 2008; Green *et al.*, 2010; see Figure 6.4).

Xylem-inhabiting pathogenic bacteria can lead to the decline and death of trees, since they block water transport and produce toxins. In Europe, certain cultivars of *Salix alba* are affected by *Erwinia salicis*, but death of the trees occurs only after they have been predisposed by other factors. Bacterial leaf scorch caused by the xylem-inhabiting bacterium *Xylella fastidiosa*, formerly limited to grape, citrus and other crops, has become a problem on landscape tree genera such as *Quercus* spp., *Ulmus* spp. and *Platanus* spp. in North America. The bacterium is transmitted by xylem-feeding leafhoppers, and it causes chronic infection leading to leaf necrosis, early leaf senescence and decline symptoms. In countries where *Eucalyptus* is native or planted, wilt caused by *Ralstonia solanacearum* has been reported, and this pathogen is the subject of quarantine regulations in many countries (Keane *et al.*, 2000; Sherald, 2007).

Apart from the use of effective antibiotics, management practices are difficult in bacterial diseases, and often limited to pruning when only parts of the tree are affected. In the future, selection of resistant or tolerant cultivars and control of insect vectors seem to be the most promising methods for dealing with many bacterial diseases.

Figure 6.4 Horse chestnut (*Aesculus hippocastanum*) dying from Pseudomonas bleeding canker and secondary infections.

6.5 Diseases caused by oomycetes

Oomycetes do not belong to the true fungi, but rather to protists called Chromista. They are often termed water moulds and are closely related to brown algae. Oomycetes reproduce by asexual sporangia that can form flagellate zoospores and are mobile in wet environments, by asexual chlamydospores that may also be airborne, and sexually by thick-walled oospores that can survive dry periods for many years. Thus, they are well adapted to many different environments, causing root infections, trunk, and stem cankers as well as leaf and needle diseases. *Phytophthora* is by far the most important genus for woody ornamentals (Erwin and Ribeiro, 1996).

Some *Phytophthora* species are generalists, with many hosts in different plant families, as is the case with *P. cinnamomi*. Together with *P. cambivora*, it is known in Europe as the cause of "ink disease" on *Castanea sativa* and a number of other hosts, but it is also responsible for extensive damage to many plant species in Australia. While *P. cinnamoni* prefers warm habitats, there are several other *Phytophthora* species with a broad host range that are relevant to urban tree health in temperate and cooler climates – for example, *P. cambivora*, *P. cactorum*, *P. syringae*, and *P. citricola* (Figure 6.5). New species continue to develop by hybridization, both in temperate and cooler regions, as well as in Australia and other warm regions (Jung *et al.*, 2011; Hayden *et al.*, 2013).

Phytophthora ramorum, the cause of Sudden Oak Death (SOD), is an internationally important quarantine organism that appeared in the 1990s. Contrary to the name of the

Figure 6.5 *Fagus sylvatica* with slime flux and cambial damage by *Phythophthora cambivora*.

disease, it has a wide host range in different plant families, including some conifers, and is thus relevant to many woody ornamentals. Most of the damage has been to *Quercus* species and *Lithocarpus densiflora* in the western coastal region of the USA, and to *Viburnum, Rhododendron* and some members of the *Fagaceae in* Europe. The pathogen also causes needle damage in plantations of Japanese larch *(Larix kaempferi)* in the UK (Rizzo *et al.,* 2002; Chadfield and Pautasso, 2012).

P. alni is an example of a *Phytophthora* species with a limited host range, occurring only on *Alnus* species. *A. glutinosa* is the primary host, but other species widely planted in urban habitats are also susceptible, such as *A. cordata* and *A.* x *spaethii.* The root system and the cambium in the lower part of the trunk are killed, leading to slime flux and, eventually, to crown dieback and death of the tree. First reports of damage to *A. glutinosa* came from Great Britain and, subsequently, the disease was found in other European countries, where it caused heavy damage to riparian ecosystems. *P. alni* is a complex of several subspecies, as a result of hybridization, some of which have been found on *Alnus* in Alaska and western Oregon. Planting stock of *Alnus* species for urban sites and the planting substrate must be free of the pathogen, and the site should not be prone to flooding (Jung and Blaschke, 2004; Ioos *et al.,* 2006).

Control and management of *Phytophthora* diseases is difficult once the pathogen is established in the ecosystem. In the case of species with a narrow host range, re-planting of the site is possible using less susceptible tree species. When pathogen species with a broad host range or quarantine species are involved, the best approach is an integrated one, consisting of a mix of sanitation and eradication procedures where feasible, application of chemicals or soil amendment and, ultimately, resistance breeding (Hayden *et al.,* 2013).

6.6 Fungal diseases

6.6.1 Systemic fungal infections

Vascular wilt diseases are the most important category of systemic infections in trees, and are responsible for some of the most devastating tree diseases observed in the past. The causal pathogens belong predominantly to the Ascomycete fungi, but there are also examples for bacterial wilts (see Section 6.4). Since the pathogens infect the water-transporting vessels in the xylem, they are, by definition, limited to angiosperms, although similar blockages of water transport in tracheids occur in gymnosperms. Symptoms consist of rapid wilting of leaves and young shoots, blockage and discolorations of vessels. Infected trees can die rapidly.

The pathogens often spread within the host vessels by conidia or yeast-like stages, and root anastomoses enable direct spread to neighboring trees. Fruit bodies form in bark or on wood, and specific or unspecific insect vectors then transmit the spores. The causal fungi and their propagules can also survive for longer periods in wood and sawdust, and enter through fresh wounds – an important fact to consider when pruning, dissecting and disposing of infected trees (Ash, 2001).

The best-documented example of a vascular tree wilt disease is Dutch Elm Disease (DED), which most likely originated in Asia and is caused by several species of fungi belonging to the *Ophiostoma ulmi* complex. The most aggressive are *O. novo-ulmi* and *O. himal-ulmi,* which developed in Eastern Europe and North America in the 1960s. Most American and European elm species are highly susceptible to the disease, whereas Asian elms are generally more resistant. The disease was first observed in Europe in the early 1900s, and was introduced to North America some time around 1929. DED has led to the

Figure 6.6 Elm selection from an early breeding program dying of Dutch elm disease.

death of many millions of elms in forests and urban areas in North America and Europe (Figure 6.6). Spread occurs primarily by bark beetles of the genus *Scolytus* that breed in dead and dying trees. Therefore, immediate removal of deadwood and diseased trees is the most important management aspect in urban areas. Elm clones with predominantly Asian parents and high resistance to Dutch elm disease are now available, based on modern breeding programs (Phillips and Burdekin, 1982; Brasier and Buck, 2002; Kirisits, 2013).

Verticillium disease is an important soil-borne vascular wilt disease caused by *Verticillium dahliae* und *V. albo-atrum*. Hundreds of woody plants, and even herbs from many different plant families, are susceptible – most of the infections in woody hosts caused by *V. dahlia*. In woody ornamentals, the fungus seems to have little host specialization. Many tree genera often used in urban areas are susceptible, including *Acer, Ailanthus, Catalpa, Fraxinus, Liriodendron, Robinia* and *Tilia*, while *Alnus, Betula, Carpinus, Fagus, Platanus* and *Populus* possess rather high resistance.

Most of the problems with Verticillium wilt are caused in *Acer* species and in the tree nurseries that produce them. The disease poses a major hygienic problem, since the fungus produces persistent microsclerotia on infected plant parts, including roots and leaves.

These can survive in soil for many years, leading to a constant increase in inoculum density and disease pressure when infected plants remain standing. Therefore, replanting affected sites requires non-susceptible species or soil replacement (Smith and Neely, 1979; Bhat and Subbarao, 1999; Tjamos *et al.*, 2000; Neubauer, 2009).

Sycamore canker, which affects *Platanus* species and the hybrid *Platanus* x *hispanica*, is caused by *Ceratocystis platani*. The fungus belongs to the *Ceratocystis fimbriata* complex, members of which affect a number of trees and agricultural crops. The name of the disease points to the fact that the pathogen, in addition to causing blue-staining and occlusion of the xylem, also leads to cankers by killing bark and cambium. Originating in eastern North America on *Platanus occidentalis*, the disease killed many *Platanus* trees in the eastern USA and in one locality in California, and was then introduced to Europe, probably during World War II. It has now spread through several European countries into Greece, where it has access to the native habitat of *Platanus orientalis*. Once established, the pathogen spreads via root contact and through pruning tools. It can survive for months on wood and on sawdust from infected sapwood, making sanitation measures difficult and costly (Engelbrecht *et al.*, 2004; Harrington, 2013).

In warmer climates, and especially in the tropics, many genera of woody plants and palms are susceptible to Fusarium wilt, caused by various races of *Fusarium oxysporum*, a soil-borne fungus that colonizes plants through rootlets and leads to blockage of water transport through the xylem. Vascular diseases caused by species of *Thielaviopsis* are also a problem in tropical tree species and in palms. For palms, heavy pruning and removal of leaf bases is not recommended, because this can promote infection by vascular pathogens (Elliott *et al.*, 2004; Gilman, 2012; and see Chapter 11).

6.6.2 Leaf and needle diseases

Most leaf diseases of broadleaved trees are caused by Ascomycetes, with the exception of foliar rust diseases (see section 6.6.4), and they cause more or less conspicuous symptoms such as small spots, large blotches or also irregular necroses. Accordingly, the terms for the diseases are *leaf spot*, *leaf blotch* and *anthracnose*. In conifers, the term *needle cast* is used if the tree species tends to shed diseased needles (e.g., *Pinus* spp.), and it is called *needle blight, brown spot*, and so on when genera are involved that tend to retain diseased needles (e.g., *Abies* spp.). Foliar and needle pathogens usually have a cycle in which the anamorphic state is produced during the growing season and the teleomorphic, or perfect, state is formed in dead tissue, or when leaves or needles have been shed. Thus, removing plant litter beneath affected trees can help to interrupt the infection cycle (Hansen and Lewis, 1997; Jones and Benson, 2001).

Some poplar clones (*Populus* spp.) are highly susceptible to Marssonina leaf spot, caused by several species of *Marssonina*. *Apiognomonia veneta*, the cause of sycamore anthracnose, leads to extensive leaf loss and also shoot blight and bark cankers on *Platanus* species (Figure 6.7). Conspicuous damage can occur following cold, wet spring weather, and repeated defoliation in the course of several years reduces host vigor (Sinclair and Lyon, 2005).

Dothistroma needle blight, caused by *Mycosphaerella pini* and its *Dothistroma* anamorph, is a needle blight pathogen with international significance. This quarantine pathogen attacks pines (*Pinus* spp.) and several other coniferous genera, and it has significance for many *Pinus* species used as ornamentals in urban settings. Bednǎřovǎ *et al.* (2013) and Bulman *et al.* (2013) treat many tree needle diseases that can also infect urban conifer plantings.

Figure 6.7 *Platanus* anthracnose (*Apiognomonia veneta*) causing death of young shoots.

6.6.3 Shoot and stem diseases and cankers

Shoot and tip blights cause more or less extensive damage to the current year's growth, sometimes progressing into older tissue. Broad-leaved trees can often respond by producing new shoots from dormant buds, whereas evergreen conifers can suffer heavy damage. *Gremmeniella abietina*, the cause of Scleroderris disease, and *Diplodia pinea*, the cause of Diplodia tip blight, belong to the most damaging conifer diseases worldwide (Figure 6.8). *Gibberella circinata*, the cause of pitch canker in many *Pinus* species, is not yet present in New Zealand and Australia, and is currently the subject of eradication efforts in Europe (Sinclair and Lyon, 2005; Capretti *et al.*, 2013; Gordon, 2013).

Ash dieback is a serious shoot disease of *Fraxinus excelsior* and other susceptible *Fraxinus* species, caused by *Hymenoscyphus fraxineus* and its *Chalara fraxinea* anamorph, which has emerged since the 1990s in Europe (Figure 6.9). The pathogen originated in Asia, but there is a basis for genetic resistance in *F. excelsior*. Ash dieback has high relevance for ash trees in urban environments, because crown deterioration and increased deadwood production pose safety concerns in susceptible ash species and cultivars (Gross *et al.*, 2014).

A multitude of Ascomycetes can cause annual or perennial cankers of the bark and cambium, represented by genera such as *Botryosphaeria, Hypoxylon, Phomopsis,* and *Sirococcus*. For many, the relationship between drought stress and disease severity has been well documented. However, there are also aggressive, often invasive, pathogens in this group. One example is chestnut blight, caused by *Cryphonectria parasitica*, which originated on Asian *Castanea* species and was introduced to North America and Europe. On both continents, it has caused damage to indigenous chestnut species, almost completely wiping out *Castanea dentata* in the eastern USA. In regions where the disease is present, urban plantings of *Castanea* spp. should be limited to resistant

Figure 6.8 *Pinus nigra* with shoot blight caused by *Diplodia pinea*.

Figure 6.9 The ash dieback pathogen *Hymenoscyphus fraxineus* can enter the main stem via lateral twigs to cause bark canker.

species, or strict disease management must be enforced (Griffin and Elkins, 1986; Hoegger *et al.*, 2003).

Butternut canker of *Juglans cinerea*, caused by *Ophiognomonia clavigignenti-juglandacearum*, is another example of a devastating canker disease. Known to be highly virulent only since 1967, it has since spread through the entire range of butternut in eastern North America. It also poses a potential threat to susceptible *Juglans* species, including *J. regia* and *J. nigra*, in other continents, should it be introduced there (Broders *et al.*, 2015).

A bark canker disease that has serious implications for road safety is Massaria disease of *Platanus* spp., caused by *Splanchnonema (Massaria) platani*. The fungus kills strips of bark, and sometimes whole branches, and subsequently leads to a soft rot in the sapwood beneath dead bark. Affected branches, even while still green, may break within months after attack, increasing the costs for tree inspection and pruning measures (LTOA, 2014).

Some bark diseases also have implications for human health, for example sooty bark disease in maple, caused by *Cryptostroma corticale*. The fungus is a saprophyte in North America and Europe that can kill trees of several *Acer* species following hot, dry summers. *C. corticale* produces abundant conidia under the periderm, which cause lung disease when inhaled. Therefore, precautions are necessary when felling and chipping maple trees suspected of having died from the disease (Gibbs, 1997).

6.6.4 Rust diseases

Rust fungi are obligate pathogens belonging to the fungal class Basidiomycetes. They lack the typical fruit bodies often associated with this class of higher fungi, but instead form lively colored sporulating structures on infected host tissue – hence the name (Figure 6.10). There are many different rust diseases on woody hosts. Many are heteroecious, that is, they have alternate hosts, often in two non-related plant genera such as a woody plant and a herb. For instance, white pine blister rust (*Cronartium ribicola*) alternates between susceptible five-needled pines (*Pinus* spp.) and *Ribes* species. It is a native European pathogen on *P. cembra*, and was accidentally introduced to North America in the late 19th century, subsequently causing heavy losses, especially among eastern white pine, *Pinus strobus*. Rust fungi are generally more important for forests, and especially plantations, but they can be relevant when hosts and alternate hosts are present in urban areas in high density (Sinclair and Lyon, 2005; Hamelin, 2013).

6.6.5 Root diseases

Oomycetes (see section 6.5), some bacteria, and fungi belonging to the Ascomycetes and Basidiomycetes, are major causes of root diseases. In some cases, fungal pathogens are also able to cause root decay and loss of wind resistance, in addition to parasitizing the root system, and they are, therefore, especially relevant to urban trees. Two important and widely distributed diseases are Armillaria root disease, present worldwide and caused by several species of *Armillaria* (Honey fungus), and Heterobasidion root and butt rot, caused by *Heterobasidion annosum* s.l., which is a pathogen group found mainly in conifer forests of the Northern hemisphere. *H. annosum* is economically more important for conifers and in forests, while *Armillaria* species can also be very damaging in urban environments. Several *Armillaria* species are able to attack commonly planted broad-leaved trees and can aggressively spread to neighboring trees via soil by their persistent rhizomorphs. Other important root pathogens include Ascomycetes belonging to the *Nectriaceae* (Guillaumin and Legrand, 2013).

Figure 6.10 Swelling and bark damage to *Juniperus communis* caused by the rust fungus *Gymnosporangium clavariaeforme*, which alternates between *Juniperus* and *Crataegus*.

6.6.6 Wood decay

Wood decay is caused by the action of specialized enzymes that are able to attack and destroy lignin, cellulose and pectins as the main components of wood. Three major decay types are usually differentiated into brown rot, white rot and soft rot, but there are transitions and variations between the types. Basidiomycetes commonly cause white and brown rots, whereas Ascomycetes are often associated with soft rots. However, some fungi can change their decay strategies and decay types, depending on conditions within the host. The development of decay in living trees has been the subject of much discussion (Schwarze *et al.*, 2000; Schmidt, 2006).

Some decay fungi are limited mainly to heartwood, such as *Laetiporus sulphureus*, whereas others like *Inonotus* and *Phellinus* species are more often associated with sapwood decay or canker rot. In urban environments, decay of the trunk or root system is one of the major reasons for removing trees or pruning them to a "safe" size. The many root injuries that urban trees sustain because of excavation activities, soil compaction and so on make root decays very common. Fungal genera such as *Ganoderma, Grifola, Kretzschmaria,* and *Meripilus* are commonly associated with root decay (Figure 6.11). Following injuries and large pruning wounds, genera like *Fomes, Laetiporus, Phellinus and Polyporus* are commonly involved in stem and trunk decay. The assessment of damage

Figure 6.11 Fruit bodies of *Ganoderma lipsiense* at the tree base, here on *Fagus sylvatica,* may point to danger of tree failure.

caused by decay and the stability of the tree and its parts against failure is difficult, and depends on many factors, such as the combination of host and fungal species, host vigor and site conditions (Schwarze, 2008; Watson and Green, 2011; and see chapter 9).

6.7 Parasitic plants

Mistletoes are hemiparasitic, chlorophyll-containing plants connected to their plant hosts by way of specialized structures called haustoria (see Chapter 10, section 10.3.10). With these, they derive water and mineral nutrients from the tree and conduct photosynthesis to produce carbohydrates. The damaging effects of mistletoes consist mainly of their mineral and, especially, water uptake from the tree, which can lead to drought stress. Dwarf mistletoes such as *Arceuthobium* spp., and leafy mistletoes such as *Viscum* spp. and *Loranthus* spp., are differentiated. Dwarf mistletoes cause heavy damage to conifer forests in western North America, but pose a threat to urban conifer species as well. They are the subject of international quarantine regulations, since there is only one indigenous species in the Old World (Parker and Riches, 1993; Shaw and Mathiasen, 2013).

In deciduous broad-leaved trees, a high rate of colonization by mistletoes of the genus *Viscum* requires a continual supply of water by the tree, increases loading by wind and snow, and may thus promote branch failure (Figure 6.12). Branch dieback distal of large mistletoes is also common. An average-size tree can support a number of perhaps 5–10 large *Viscum album* mistletoes without sustaining damage. Therefore, in urban areas,

Figure 6.12 Infestation of silver maple (*Acer saccharinum*) by mistletoe (*Viscum album*) associated with branch failure at the top centre of tree crown.

a balance should be sought between mistletoes – which are broadly viewed as ecologically valuable – and the well-being of the tree.

Plant-parasitic weed species belonging to genera such as *Cuscuta* or *Orobanche* also parasitize woody plants, attaching to aboveground parts, or to roots in some cases. Parasitism of woody hosts by species of *Cuscuta* can greatly damage the host plants, especially in warmer climates and in the tropics (Parker and Riches, 1993; Sinclair *et al.*, 2005).

6.8 Plant-parasitic nematodes and insect pests

Plant-parasitic nematodes feed on plant cells and damage roots, leaves and needles, as well as the xylem tissue, where blockage of water can occur. The pine wood nematode (PWN, *Bursaphelenchus xylophilus*), native to North American conifer species, is now the subject of quarantine regulations throughout the world. In Japan and Portugal, PWN has killed many forest and amenity trees, and *Pinus* and other conifer species in Europe, including those used in urban settings are at risk if it spreads (Kamata and Takeuchi, 2013).

Figure 6.13 Infestation by the horse chestnut leaf miner (*Cameraria ohridella*) leads to a reduction in aesthetic value.

Insects that are damaging to trees and shrubs are so multitudinous that it is not possible to treat them in detail here. Contrary to forest ecosystems, the main focus in urban areas is on managing insect damage to individual trees and shrubs, and estimating the influence on aesthetics and physiology of the host before possibly taking counter-measures (Alford, 1991; Johnson and Lyon, 1991; Davidson *et al.*, 2014).

There are several examples of widespread pests relevant to popular urban tree species. The horse chestnut leaf miner (*Cameraria ohridella*) causes leaf damage to *Aesculus hippocastanum*, and has spread throughout Europe and to Scandinavia rapidly since it first appeared in Macedonia in 1986 (Figure 6.13). At first viewed to be an invasive species, it is now known to be native to the southern Balkan region (Lees *et al.*, 2011).

The sycamore lace bug (*Corythucha ciliata*) is native to North America, and was introduced to Italy in the early 1960s. It reduces aesthetic value of *Platanus* species and has since spread to many countries in Central and Western Europe and the UK (Figure 6.14). Another insect with a broad host range in many broadleaved urban tree species, the horse chestnut scale (*Pulvinaria regalis*) causes sucking damage to leaves and young twigs. Presumably introduced to the UK from Asia in the 1960s, it has since spread to several countries in mainland Europe. However, the population seems to fluctuate quite heavily, due to natural antagonists, enabling feasible approaches to biological control (Arnold and Segonca, 2003; Watson, 2013).

Buprestid beetles, bark beetles, bark-feeding weevils, defoliating caterpillars, wood-boring caterpillars of butterflies and moths (Figure 6.15), as well as wood-boring beetle larvae, can inflict heavy damage and kill trees, especially freshly planted ones and those predisposed by various abiotic factors.

Presently, some invasive insect species of international concern directly affect urban areas, and are regulated as quarantine pests in various parts of the world where they are not yet present or established. Monitoring and management of these pests is necessary in

Figure 6.14 Sycamore lace bug (*Corythucha ciliata*) leads to yellowing of foliage and reduction in aesthetic value, here in *Platanus* x *hispanica*.

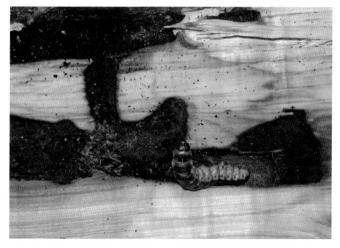

Figure 6.15 Wood-boring caterpillars of the goat moth (*Cossus cossus*) produce large galleries in sapwood – here in *Salix caprea* – leading to mechanical weakening of stems and branches.

urban tree populations. Emerald ash borer (EAB, *Agrilus planipennis*) originates in Asia and was introduced to North America, where it has caused extensive damage to *Fraxinus* species in urban areas and forests. Infestation by EAB also structurally weakens branches, and can promote branch failure. The Asian longhorned beetle (ALB, *Anoplophora glabripennis*) was introduced to the New York region around 1996, and to Austria and Germany in 2001. Potential for damage to urban trees is high in Europe as well as in the USA. The large feeding galleries can lead to branch death and failure, and thus ALB also poses a threat to tree safety (Nowak *et al.*, 2001; Poland and McCullough, 2006).

Some insects of urban trees have a direct impact on human or animal health in urban environments. One example is oak processionary moth (*Thaumatopoea processionae*), which colonizes indigenous and exotic *Quercus* species in Europe. The nest-building larvae feed on leaves, and form urticating hairs that not only lead to skin infections, conjunctivitis and asthmatic complaints in humans, but also in pets and other animals.

6.9 Damage by herbivorous mammals

Where high population levels of rodents such as rabbits or voles are present, damage, especially to young and thin-barked trees, can occur. Voles of the genus *Microtis* girdle the bark at the tree base or feed on roots, leading to the death of young trees. Grey squirrels, introduced to Great Britain over 100 years ago, cause bark-stripping damage in urban woodlands. Pasture animals like sheep, goats and horses, as well as ungulate wildlife such as deer, can cause heavy browsing and bark-stripping damage, for example along rural roads and in less densely populated urban areas. Where these problems are present, it is advised to protect the stems of young plants, which also serves as protection against sunscald (Salmon *et al.*, 2006).

6.10 Impact of introduced pests and diseases

In the past century, invasive pests and pathogens have caused many of the most damaging tree diseases. Studies have attempted to estimate the cost of invasive species to agriculture and forestry. Most such studies focus on the damage caused to forests, but the Dutch elm disease pandemic shows the risk involved in urban environments when one tree species makes up a high proportion of amenity trees. Urban tree managers must be aware that, in addition to the direct costs arising from loss of amenity trees and ornamentals, dealing with invasive pests and pathogens also requires administration capacity. ALB, PWN and SOD are good examples where extraordinary measures are necessary when invasive species are detected (National Research Council, 2002; Rizzo *et al.*, 2002).

6.11 Aspects of control methods for pests and diseases of urban trees

In principle, it is possible to control most diseases and pests using integrated protection methods that combine cultural practices, biological, chemical and genetic approaches. In the past, chemical methods were often relied upon. For instance, systemic fungal diseases of trees can be treated by injecting fungicides. However, treatments must be repeated, and stem injection wounds can cause damage to the tree. Pesticides are still heavily used in urban trees in cases where quarantine pests must be regulated – for instance, stem injection of imidacloprid in combating the Asian longhorned beetle (ALB) and the emerald ash borer (EAB) in the USA. The detrimental effects of such treatments, such as increased infestation by other pests, as well as concerns regarding effects on soil and non-target organisms, must be taken seriously (Kreutzweiser *et al.*, 2008; Mota-Sanchez *et al.*, 2009).

In many urban areas, public acceptance of chemical plant protection methods has decreased dramatically so that, in some communities, the use of these substances, though feasible, is highly restricted. Instead, the focus is on developing integrated approaches that include cultural and biological methods. For instance, hypovirulence (i.e., reduced virulence caused by viral infection of the pathogen) can be introduced artificially into pathogen populations of the chestnut blight fungus, *Cryphonectria parasitica* (Hoegger *et al.*, 2003).

In the long term, resistance breeding and the selection of resistant and tolerant tree species and cultivars must be the goal, but this is an ongoing process, since pests and diseases also adapt and can overcome host resistance.

6.12 Conclusions

Diseases, pests and disorders of amenity trees and woody ornamentals are a natural part of the urban environment. The presence of a multitude of weak and secondary pathogens and pests is the reason why it is so important to maintain trees, especially freshly planted ones, in a vigorous condition. Conversely, damage by secondary pests and pathogens is a valuable indication that the plant is either not suitable for the site, or has not been cared for properly.

A major challenge in urban areas is to keep ancient and veteran trees in a safe condition in spite of multiple defects caused by decay fungi. This requires management concepts that combine aesthetics with appropriate tree care methods and safety considerations, and enable communities to preserve exceptional, old trees (Lonsdale, 2013).

Many urban environments depend on only a few species and cultivars of amenity trees and ornamentals, making them vulnerable to new pests and diseases and to the effects of climate change. Knowledge of local species composition and vulnerability to pests is valuable knowledge for communities (Vecht and Conway, 2015).

The distribution of risk among many different plant species, and the establishment of site-adequate, well-maintained and, thus, vigorous plantings is an important prerequisite for keeping urban green spaces healthy and is a key challenge for the future (see Chapter 9).

References I: Recommended diagnostic literature

Alford, D.V. (1991). *A color atlas of pests of ornamental trees, shrubs and flowers*. Wolfe Publishing Ltd., London.

Bond, J. (2012). *Urban Tree Health. A practical and precise estimation method*. Urban Forest Analytics, Geneva NY.

Butin, H., Nienhaus, F. and Böhmer, B. (2010). *Farbatlas Gehölzkrankheiten*. Ulmer, Stuttgart.

Chase, A.R. and Broschat, T.K. (eds, 1991). *Diseases and disorders of ornamental palms*. APS Press, St. Paul, Minnesota.

Costello, L.R., Perry, E.J., Matheny, N.P., Henry, J.M. and Geisel, P.M. (2003). *Abiotic disorders of landscape plants*. University of California Agriculture and Natural Resources Communications Services, Richmond, California.

Davidson, J.A. and Raupp, M.J. (2014). *Managing insects and mites on woody plants: an IPM approach*, 3rd edition. Tree Care Industry Association, Londonderry, New Hampshire.

Dreistadt, S.H., Clark, J.K. and Flint, M.L. (2004). *Pests of landscape trees and shrubs*, 2nd edition. University of California Agriculture and Natural Resources Publication 3359.

Elliott, M.L., Broschat, T.K., Uchida, J.Y. and Simone, G.W. (eds, 2004). *Compendium of Ornamental Palm Diseases and Disorders*. APS Press, St. Paul, Minnesota.

Hansen, E.M. and Lewis, K.J. (1997). *Compendium of conifer diseases*. APS Press, St. Paul, Minnesota.

Johnson, W.T. and Lyon, H.H. (1991). *Insects that feed on trees and shrubs*, 2nd edition. Cornell University Press, Ithaca, New York.

Jones, R.K. and Benson, D.M. (eds). (2001). *Diseases of woody ornamentals and trees in nurseries*. APS Press, St. Paul, Minnesota.

Keane, P.J., Kile, G.A., Podger, F.D. and Brown, B.N. (2000). *Diseases and Pathogens of Eucalypts*. CSIRO Publishing, Clayton, Victoria, Australia.

Ogawa, J.M., Zehr, E.I., Bird, G.W., Ritchie, D.F., Uriu, K. and Uyemoto, J.K. (1995). *Compendium of stone fruit diseases*. APS Press, St. Paul, Minnesota.

Old, K. M., Wingfield M.J. and Yuan, Z.Q. (2003). *A manual of diseases of eucalypts in South-East Asia*. Center for International Forestry Research, Jakarta.

Ozanne, C. (2010). Insect pests of some important forest trees. In: Bowes, B.G.B. (ed). *Trees and forests, a colour guide*, pp. 178–197. Manson Publishing, London.

Sinclair, W.A. and Lyon, H.W. (2005). Diseases of trees and shrubs. 2nd ed. Cornell University Press, Ithaca.

Watson, G. (2013). *Tree Pests and Diseases. An Arborists' Field Guide*. The Arboricultural Association, Gloucestershire.

Watson, G. and Green, T. (2011). *Fungi on trees. An Arborists' Field Guide*. The Arboricultural Association, Gloucestershire.

Woodward, S. (2010). Microbial and viral pathogens, and plant parasites of plantation and forest trees. In: Bowes, B.G.B. (ed). *Trees and forests, a colour guide*, pp. 139–177. Manson Publishing, London.

References II

Arnold, C. and Sengonca, C. (2003). Possibilities of biological control of the horse chestnut scale insect, *Pulvinaria regalis* Canard (Homoptera: Coccidae), on ornamental trees by releasing its natural enemies. *Journal of Plant Diseases and Protection* **110**, 591–601.

Ash, C.L. (ed, 2001). *Shade Tree Wilt Diseases*. APS Press, St. Paul, Minnesota.

Bednářová, M, Dvořák, M., Janousek, J. and Jankowsky, L. (2013). Other foliar diseases of coniferous trees. In: Gonthier, P., Nicolotti, G. (eds). *Infectious forest diseases*, pp. 458–487. CAB International, Oxfordshire.

Bhat, R.G. and Subbarao, K.V. (1999). Host range specificity in *Verticillium dahliae*. *Phytopathology* **89**, 1218–1225.

Brasier, C.M. and Buck, K.W. (2002). Rapid evolutionary changes in a globally invading fungal pathogen (Dutch Elm Disease). *Biological Invasions* **3**, 223–233.

Broders, K., Boraks, A., Barbison, L., Brown, J. and Boland, G.J. (2015). Recent insights into the pandemic disease butternut canker caused by the invasive pathogen *Ophiognomonia clavigignenti-juglandacearum*. *Forest Pathology* **45**, 1–8. doi: 10.1111/efp.12161.

Bulman, L.S., Dick, M.A., Ganley, R.J., McDougal, R.L., Schwelm, A. and Bradshaw, R. (2013). Dothistroma needle blight. In: Gonthier, P., Nicolotti, G. (eds). *Infectious forest diseases*, pp. 436–457. CAB International, Oxfordshire.

Butin, H. (1995). *Tree diseases and disorders*. Oxford University Press, New York/Tokyo.

Büttner, C., von Bargen, S., Bandte, M. and Mühlbach, H.M. (2013). Forest diseases caused by viruses. In: Gonthier, P. and Nicolotti, G. (eds). *Infectious forest diseases*, pp. 50–75. CAB International, Oxfordshire.

Capretti, P., Santini, A. and Solheim, H. (2013). Branch and tip blights. In: Gonthier, P. and Nicolotti, G. (eds). *Infectious forest diseases*, pp. 420–435. CAB International, Oxfordshire.

Chadfield, V. and Pautasso, M. (2012). *Phytophthora ramorum* in England and Wales: which environmental variables predict county disease incidence? *Forest Pathology* **42**, 150–159.

Cooper, J.I. (1993). *Virus Diseases of Trees and Shrubs*. Chapman & Hall, London.

Desprez-Loustau, M.L., Marcais, B., Nageleisen, L-M., Piou, D. and Vannini, A. (2006). Interactive effects of drought and pathogens in forest trees. *Annals of Forest Science* **63**, 597–612.

Engelbrecht, C.J.B, Harrington, T.C., Steimel, J. and Capretti, P. (2004). Genetic variation in eastern North American and putatively introduced populations of *Ceratocystis fimbriata f. platani. Molecular Ecology* **13**, 2995–3005.

Erwin, D.C. and Ribeiro, O.K. (1996). *Phytophthora Diseases Worldwide*. APS Press, St. Paul, Minnesota.

Gibbs, J.N. (1997). Fifty years of sooty bark disease of sycamore. *Quarterly Journal of Forestry* **91**, 215–221.

Gilman, E.F. (2012). *An illustrated guide to pruning*, 3rd edition. Cengage Learning, Clifton Park, New York.

Gordon T.R. (2013). Pitch Canker. In: Gonthier, P. and Nicolotti, G. (eds). *Infectious forest diseases*, pp. 376–391. CAB International, Oxfordshire.

Green, S., Studholme, D.J., Laue, B., Dorati, F., Lovell, H., Arnold, D., Cottrell, J.E., Bridgett, S., Blaxter, M., Huitema, E., Thwaites, R., Sharp, P.M., Jackson, R.W. and Kamoun, S. (2010). Comparative genome analysis provides insights into the evolution and adaptation of *Pseudomonas syringae pv. aesculi* on *Aesculus hippocastanum. PloS One* **5**, e10224, 1–14.

Griffin, G.J. and Elkins, J.R. (1986). Chestnut blight. In: Roane, M. *et al.* (eds). *Chestnut blight, other Endothia diseases and the Genus Endothia*, pp. 1–26. APS Mongraph Series, St.Paul, Minnesota.

Griffiths, H.M. (2013). Forest diseases caused by Prokaryotes: phytoplasmal and bacterial diseases. In: Gonthier, P. and Nicolotti, G. (eds). *Infectious forest diseases*, pp. 76–96. CAB International, Oxfordshire.

Gross, A., Holdenrieder, O., Pautasso, M., Queloz, V. and Sieber, T.N. (2014). Hymenoscyphus pseudoalbidus, the causal agent of European ash dieback. *Molecular Plant Pathology* **15**, 5–21.

Guillaumin, J.-J. and Legrand, P. (2013). Armillaria root rots. In: Gonthier, P. and Nicolotti, G. (eds). *Infectious forest diseases*, pp. 159–177. CAB International, Oxfordshire.

Hamelin, R.C. (2013). Tree rusts. In: Gonthier, P. and Nicolotti, G. (eds). *Infectious forest diseases*, pp. 547–566. CAB International, Oxfordshire.

Harrington, T.C. (2013). Ceratocystis disease. In: Gonthier, P. and Nicolotti, G. (eds). *Infectious forest diseases*, pp. 230–255. CAB International, Oxfordshire.

Harris, R.W., Clark, J.R. and Matheny, N. (2004). *Arboriculture – Care of landscape trees, shrubs and vines*, 4th edition, Prentice Hall Inc., Upper Saddle River, New Jersey.

Harrison, N.A. and Jones, P. (2004). Lethal yellowing. In: Elliott, M.L., Broschat, T.K., Uchida, J.Y. and Simone, G.W. (eds). *Compendium of Ornamental Palm Diseases and Disorders*, pp. 39–4. APS Press, St. Paul, Minnesota1.

Hayden, K.J., Hardy, G.E. and Garbelotto, M. (2013). Oomycete diseases. In: Gonthier, P. and Nicolotti, G. (eds). *Infectious forest diseases*, pp 519–546. CAB International, Oxfordshire.

Hepting, G.H. (1974). Death of the American Chestnut. *Journal of Forest History* **18**, 60–67.

Hoegger, P.J., Heiniger, U., Holdenrieder, O. and Rigling, D. (2003). Differential transfer and dissemination of hypovirus and nuclear and mitochondrial genomes of a hypovirus-infected *Cryphonectria parasitica* strain after introduction into a natural population. *Applied and Environmental Microbiology* **69**, 3767–3771.

Houston, D.R. (1994). Major new tree disease epidemics: Beech bark disease. *Annual Review of Phytopathology* **32**, 75–87.

Ioos, R., Andrieux, A., Marcais, B. and Frey, P. (2006). Genetic characterization of the natural hybrid species *Phytophthora alni* as inferred from nuclear and mitochondrial DNA analyses. *Fungal Genetics and Biology* **43**, 511–529.

Jung, T. and Blaschke, M. (2004). Phytophthora root and collar rot of alders in Bavaria: distribution, modes of spread and possible management strategies. *Plant Pathology* **53**, 197–208.

Jung, T., Stukely, M.J.C., Hardy, G.E., White, D., Paap, T., Dunstan, W.A. and Burgess, T.I. (2011). Multiple new *Phytophthora* species from IST clade 6 associated with natural ecosystems in Australia: evolutionary and ecological implication. *Persoonia* **26**, 13–39.

Kamata, N. and Takeuchi, Y. (2013). Pine wilt disease and other nematode diseases. In: Gonthier, P. and Nicolotti, G. (eds). *Infectious forest diseases*, pp. 115–127. CAB International, Oxfordshire.

Kirisits, T (2013). Dutch Elm Disease and other Ophiostoma diseases. In: Gonthier, P. and Nicolotti, G. (eds). *Infectious forest diseases*, pp. 256–282. CAB International, Oxfordshire.

Kreutzweiser, D.P., Good, K.P., Chartrand, D.T. Scarr, T.A. and Thompson, D.G. (2008). Are leaves that fall from imidacloprid-treated maple trees to control Asian longhorned beetles toxic to non-target decomposer organisms? *Journal of Environmental Quality* **37**, 639–646.

Lee, I.-M., Davis, R.E. and Gundersen-Rindal, D.E. (2000). Phytoplasma: phytopathogenic mollicutes. *Annual Review of Microbiology* **54**, 221–255.

Lees, D. C., Lack, H. W., Rougerie, R., Hernandez-Lopez, A., Raus, T., Avtzis, N., Augustin, S. and Lopez-Vaamonde, C. (2011). Tracking origins of invasive herbivores through herbaria and archival DNA: the case of the horse-chestnut leaf miner. *Frontiers in Ecology and the Environment* **9**, 322–328.

Lonsdale, D. (ed, 2013). *Ancient and other veteran trees: further guide on management.* The Tree Council, London.

LTOA (2014). *Massaria disease of plane: practical management guidance.* London Tree Officers Association (eds.), London, 28 pp.

Manion, P.D. (1981). *Tree disease concepts.* Prentice-Hall, Englewood Cliffs, N.J.

Manion, P.D. and Lachance, D. (1992). *Forest Decline Concepts.* APS Press., St. Paul, Minnesota.

Mota-Sanchez, D., Cregg, B.M., McCullough, D.G., Poland. T.M. and Hollingworth, R.M. (2009). Distribution of trunk-injected 14C-imidacloprid in ash trees and effects on emerald ash borer (Coleoptera: Buprestidae) adults. *Crop Protection* **28**, 655–661.

National Research Council (eds, 2002). *Predicting Invasions of Nonindigenous Plants and Plant Pests: Committee on the Scientific Basis for Predicting the Invasive Potential of Nonindigenous Plants and Plant Pests in the United States.* National Academy Press, Washington D.C.

Neubauer, C., Heitmann, B. and Vogel, C. (2009). Morphology, vegetative compatibility and pathogenicity of *Verticillium dahliae* isolates from woody ornamentals in Germany. *Journal of Plant Diseases and Protection* **116**, 109–114.

Nowak, D.J., Pasek, J.E., Sequeira, R.H., Crance, D.E. and Mastro, V.C. (2001). Potential effect of *Anoplophora glabripennis* (Coleoptera: Cerambycidae) on urban trees in the United States. *Forest Entomology* **94**, 116–122.

Parker, C. and Riches, C.R. (1993). *Parasitic weeds of the world: biology and control.* CAB International, Wallingford, UK.

Phillips, D.H. and Burdekin, D.A. (1982). *Diseases of forest and ornamental trees.* MacMillan Press, London.

Poland, T.M. and McCullough, D.G. (2006). Emerald ash borer: invasion of the urban forest and the threat to North America´s ash resource. *Journal of Forestry* **104**, 118–124.

Rizzo, D.M., Garbelotto, M., Davidson, J.M., Slaughter, G.W. and Koike, S.T. (2002). *Phytophthora ramorum* as the cause of extensive mortality of Quercus spp. and *Lithocarpus densiflorus* in California. *Plant Disease* **86**, 205–214.

Salmon, T.P., Whisson, D.A. and Marsh, R.E. (2006). *Wildlife pest control around gardens and homes.* University of California Agriculture and Natural Resources Publications 21385, Oakland, California.

Schink, B., Ward, J.C. and Zeikus, J.G (1981). Microbiology of wetwood: importance of pectin degradation and Clostridium species in living trees. *Applied and Environmental Microbiology* **42**, 526–532.

Schmidt, O. (2006). *Wood and tree fungi: biology, damage, protection and use.* Springer, Berlin.

Schwarze, F.W.M.R. (2008). *Diagnosis and prognosis of the development of wood decay in urban trees.* ENSPEC Pty Ltd., Rowville, Australia.

Schwarze, F.W.M.R., Engels, J. and Mattheck, C. (2000). *Fungal Strategies of Wood Decay in Trees.* Springer, Berlin.

Shaw, D.C. and Mathiasen, R.L. (2013). Forest diseases caused by higher parasitic plants: mistletoes. In: Gonthier, P. and Nicolotti, G. (eds). *Infectious forest diseases*, pp. 97–114. CAB International, Oxfordshire.

Sherald, J.J. (2007). Bacterial leaf scorch of landscape trees: what we know and what we do not know. *Arboriculture and Urban Forestry* **33**, 376–385.

Smith, L.D. and Neely, D. (1979). Relative susceptibility of tree species to *Verticillium dahliae*. *Plant Disease Report* **63**, 328–332.

Stipes, R.J. (1999). Fusarium wilt of trees. In: Ash, C.L. (ed) *Shade Tree Wilt Diseases*, pp.165–169. APS Press, St. Paul, Minnesota.

Tainter, F.H. and Baker, F.A. (1996). *Principles of Forest Pathology.* John Wiley, New York.

Tattar, T.A. (1978). *Diseases of shade trees.* Academic Press, New York.

Tjamos, E.C., Rowe, R.C., Heale, J.B. and Fravel, D.R. (eds, 2000). *Recent advances in Verticillium research and management.* APS Press, St. Paul, Minnesota.

Vanneste, J.L. (ed, 2000). *Fire blight: The disease and its causative agent, Erwinia amylovora*. CABI Publishing, Wallingford.

Vecht, J.V. and Conway, T.M. (2015). Comparing species composition and planting trends: exploring pest vulnerability in Toronto's urban forest. *Arboriculture and Urban Forestry* **41**, 26–40.

Webber, J., Parkinson, N., Rose, J., Stanford, H., Cook, R.T.A. and Elphinstone, J. (2008). Isolation and identification of *P. syringae pv. aesculi* causing bleeding canker of horse chestnut in the UK. *Plant Pathology* **57**, 368.

CHAPTER 7

Vitality assessment, tree architecture

Andreas Roloff

Technische Universität Dresden, Tharandt, Germany

7.1 Introduction

In this chapter, general aspects of tree architecture and methods of tree vitality assessments are discussed, and existing disparities or contradictions are pointed out when assessments based on "leaf loss" and based on crown structures are compared. The necessity of considering the branching pattern is substantiated, and the methods developed to date are presented.

It is still very difficult to determine tree vitality and, thereby, the effects of stress and decline in deciduous trees. The reason for this difficulty is due to the fact that, until now, most inventories considered only parameters such as "percentage leaf loss". Therefore, this chapter puts focus on the branching pattern and the crown structure of trees in the assessment of tree vitality.

7.2 Decline and stress symptoms of tree crowns: "leaf loss" vs. crown structure

The consideration of crown structures in the assessment of tree vitality has become increasingly important. That scientists are now aware of the problem of "leaf loss" has been shown in many studies (Roloff, 1989). The number of leaves, and above all the leaf size, is subject to considerable annual fluctuations – for example, as a result of drought and insect damage or flowering and fructification. However, leaf size can vary greatly, even within the same crown of a deciduous tree. Thus, it is difficult to show a statistical significance in values between different trees. The correlation between "leaf loss" and fructification intensity has been well demonstrated but, on the other hand, the foliage must also be considered if possible.

In this context, there is one more aspect which should be mentioned, for it has become particularly apparent in the most recent investigations. There is a great variety of tree species in which, with increasing shoot lengths (i.e., with better growth), the crown becomes more transparent. In this case, a vitality assessment on the basis of crown transparency versus crown structure is bound to produce exactly opposite results (Roloff, 1989). Finally, there are pioneer species, such as many birch and pine species, which

---

Urban Tree Management: For the Sustainable Development of Green Cities, First Edition. Edited by Andreas Roloff.
© 2016 John Wiley & Sons, Ltd. Published 2016 by John Wiley & Sons, Ltd.

never show a crown without any gap or transparency, because this is typical for the species strategy – the light crown.

For this reason, it is not surprising to find considerable disagreement between so-called "leaf loss" and crown structure (to be discussed in the following chapter), which occurs when vigor assessments of the same trees are compared. Assessments may differ markedly, and agreement is only achieved in about 50% of the assessed trees.

Therefore, it would be advantageous if the term "leaf loss" were to be replaced by a different, more objective term such as "crown transparency", which does not lead to the misconception of shedded leaves. A deciduous tree showing a "leaf loss" of 30% does not mean that 30% of the leaves have been shed. These leaves often simply never existed at the beginning of the growing season because of gaps in the branches.

7.3 Tree architecture and reiterations

7.3.1 Architectural models

On the basis of so-called architectural models, woody species (trees and shrubs) all over the world can be classified into types with similar branching patterns and crown development (Hallé *et al.*, 1978; Oldeman, 2014). Important criteria are the direction in which top shoots and lateral shoots grow, the length of the growth period and the position of the flowers. The relevant models, named after renowned botanists (e.g., Rauh), will be introduced in the following. Considering that 23 models can be differentiated and described worldwide, the number of the most important ones for trees can be reduced to five due to the following facts:
• the *trunk is ramified* (this is not the case in several species of palm trees);
• the *vegetative axes are differentiated* into a dominant trunk and subordinate branches (which is usually not the case in shrubs);
• most woody plants *grow rhythmically*, with clearly discernible annual growth modules due to dormancy in winter or during drought periods.

The abovementioned five most important architectural models are defined by the following characteristics (see Figure 7.1):
• **Rauh**: All shoots are ± vertically oriented, flowers in lateral position.
• **Scarrone**: All shoots are ± vertically oriented, flowers in terminal position.
• **Massart**: Vertical trunk, ± horizontal lateral branches.
• **Champagnat**: All top shoots grow vertically at first; some become horizontal by turning downwards secondarily.
• **Troll**: All shoots grow horizontally at first; the top shoot straightens up in a secondary stage.

Certain limitations have to be considered in the application of the architectural models. For instance, a number of studies have shown that many species do not belong only to a single architectural model, either because they do not meet the criteria clearly enough, or because a tree may change models in the course of its life, or due to environmental influences. The hypothesis that the architectural models represent and are connected to various ecological adaptations has proved right in some cases (e.g., in the model "Troll" – shade-tolerant species).

However, the concept of classifying trees on the basis of architectural models is still a decisive help in categorizing the great variety of crowns in trees and shrubs, and in making them easily accessible for further studies.

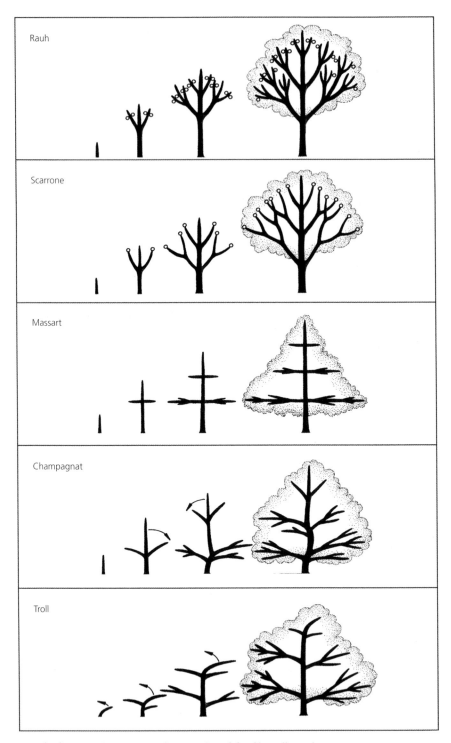

Figure 7.1 The five most important architectural models of broadleaved trees.

7.3.2 Reiterations

After the "genetic program" of trees has been described using architectural models, the following will concern themselves with variations on the inherited model and with environmental influences.

A young tree can usually only grow according to its architectural model for a very short time (if at all) before it has to react to environmental influences such as loss of the terminal bud or even of the entire leader shoot (e.g., due to vandalism – see Figure 7.2), sudden changes in the lighting or the spatial situation (e.g., if neighboring trees are removed), droughts, urban stress, flood, and so on.

In order to react to these influences, the tree can no longer rely on its species-specific growth program. In many tree species, the most important basis for reactions to environmental influences are so-called *reiterations* – the development of new shoots or axes, and even entire branching systems, which reiterate the tree's architectural model, but whose development was not to be expected where and when it occurred.

The term "reiteration" is not identical with "repetition". It applies only to unexpected repetitions of the entire architectural model. On the other hand, reiterations are more than simple regenerations or "repair mechanisms". They are not only caused by stress and damages, but also by optimum or suddenly improved growth conditions.

Reiterations cannot be predicted from the architectural model. They result in the development of smaller sub-crowns within the actual crown, which are discernible as smaller trees which grow in the crown of the "mother tree" (Figure 7.3).

The concept of reiterations facilitates a systematic categorization of a tree's reactions to environmental influences, which is crucial in understanding and studying the crown structures of older trees. As a rule, the number of reiterations increases with age.

Figure 7.2 Leader damage by vandalism and compensation by reiterations (*Fraxinus excelsior*).

Figure 7.3 Different types of reiteration in an older tree, discernible by smaller upright sub-crowns.

A very helpful classification distinguishes three different "sources" of reiterations:
- sprouting of *dormant buds*;
- development and sprouting of (newly formed) *adventitious buds* (e.g., from roots or wound tissue);
- re-orientation of lateral branches from horizontal to vertical growth.

Independent of their origin, reiterations can have two different causes:
- *Traumatic reiterations* are caused by sudden, negative environmental influences, especially by injuries (Figure 7.4).
- *Adaptive reiterations* are caused by changes in the environment or the surroundings (e.g., a change in lighting conditions when neighboring trees are removed).

Finally, species can be divided into three groups, according to their reiterative potential:
- species with *no capacity* for reiterations at all (only some species of tropical palm trees);
- species which are capable *only* of *traumatic reiterations* (various species of palm trees);
- species with both *traumatic* and *adaptive* reiterations (all other woody plants).

Our own studies have shown that a great variety of possible reiterations can be observed, even though most of them occur infrequently in most species. Another common way to react to changing environmental influences is varying the shoot lengths (see following paragraphs).

Figure 7.4 Traumatic reiteration: secondary crown development after severe root injury caused by road construction work (*Tilia cordata*).

7.4 Changes in the crown structure with decreasing vitality

In this section, tree vitality is discussed in terms of growth potential which, in trees, is expressed in shoot growth. Although various branching structures within one tree crown have been known for a long time (see, for example, Büsgen & Muench, 1929; Thiebaut, 1988), their significance as a vitality indicator has only recently been discovered (Roloff, 1989, 2004).

7.4.1 Shoot morphology: shoot base scars, short- and long-shoots

By careful observation of the branching pattern of a hardwood tree, closely packed grooves upon the shoot surface can be recognized conspicuously (Figure 7.5). The significance of these shoot-base scars has, unfortunately, been ignored for a long time. They are the scars of the bud scales which, when closely packed, originally encased the young shoot primordia, and hence mark the boundary between two years of growth exactly to a millimeter. Thus, it becomes possible to retrace the development of any temperate hardwood tree branching pattern for many years (in some species such as beech, over decades), and to reconstruct its growth in this way.

Further investigating the branching pattern in most broad-leaved tree species can distinguish two kinds of shoots: short-shoots and long-shoots (Figure 7.5). Short-shoots are only a few millimeters or centimeters long, have only 3–5 leaves, and do not ramify during the following years because they only bear small dormant lateral buds. The terminal

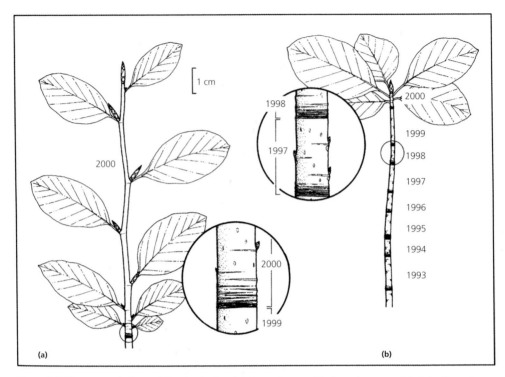

Figure 7.5 Shoot morphology of *Fagus sylvatica*. **(a)** Long-shoot with shoot base scar (circle), alternate distichous leaves and lateral buds, terminal bud. **(b)** a nine year old short-shoot chain without any lateral twigs, and with a terminal cluster of leaves.

bud of short-shoots, however, either produces a short-shoot again the following season so that short-shoot chains are formed, or else it returns to forming a long-shoot. Long-shoots are clearly longer, show more leaves, and ramify during the following year.

In any tree species, the annual growth length of the treetop shoots, and therefore the height increment of the tree, decreases after passing a culmination point, because of maturity. This finding reflects decreasing vigor in the tree. The lengths of treetop shoots are interpreted as a sign of vitality, because the strategy of a forest-forming tree species which strives to conquer new airspace must occur at the very top of the tree. On the other hand, the shoot lengths in the inner, lower, and lateral crown areas mainly depend on competition and light conditions, and are therefore unsuitable for the vigor assessment of a whole tree.

7.4.2 Model of growth stages

In every investigated (broad-leaved) tree species, there are four growth stages to discriminate: exploration, degeneration, stagnation, and retraction. These stages, which result in fundamental modifications of the branching structure, are due to (statistically significant) decreasing annual shoot lengths. Especially in the leafless state, these different branching structures in the treetop are easily to perceive, even from a distance (and in aerial photographs, too). They are the basis of vitality assessment in four vitality classes.

Figure 7.6 shows how a typical branching pattern originates from a leader shoot of a vigorous beech tree. This branching pattern is similar in most other temperate hardwoods,

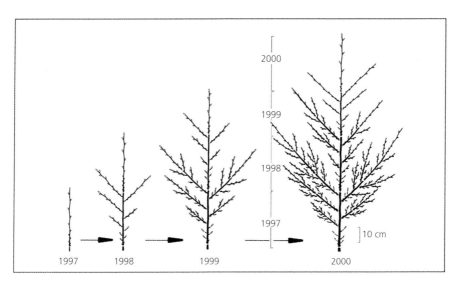

Figure 7.6 Branching pattern of a vigorous hardwood tree leader shoot during three years' development (*Fagus sylvatica*).

with only few modifications. This "*exploration phase*" produces the branching structure that is known best and found most frequently. The terminal and upper lateral buds yearly develop long-shoots, the lower lateral buds develop short-shoots and, finally, the lowest lateral buds do not shoot at all, but remain as very small dormant buds for years, preserved for unusual events (preventitious buds).

On every annual leader shoot, the lengths of the younger lateral shoots decrease from top to bottom and the developing branching pattern is turned upwards or forwards (acrotony). In this way, an obviously storied branch system is developed, and the annual shoot boundaries (marked by interruptions of the black lines in Figure 7.7) can be distinguished even from a distance, by the steps of the branching pattern and by the abrupt change of long lateral shoots to short-shoots. This exploration phase is the widespread appearance of the leader shoots in healthy vigorous trees until old age, because this is the only way the treetop can fulfil its main purpose for the tree – namely, to conquer new airspace steadily, to fill it up with lateral shoots, and to be successful against competing trees.

In the "*degeneration phase*" (Figure 7.7), however, the terminal bud develops shorter long-shoots but, from nearly any lateral bud (including the uppermost), short-shoots arise almost without exception. Therefore, an obvious impoverishment of the branching pattern takes place, and spears ("fox-tails") are formed in the periphery of the crowns, which may also be seen from a great distance.

In the course of further decreasing vitality, even the terminal bud changes into developing short-shoots. In this "*stagnation phase*", ramification ceases, because short-shoots do not ramify. Because of the short annual length of these shoots the length increment of the branch and the height increment of the tree, respectively, stagnates.

If this stagnation phase persists longer than a few years (if it is not only temporary), the branch or (if the treetop shoots are concerned) the treetop dies back. This stage is called the "*retraction phase*".

As a result of their disadvantageous mechanical static features (a dense cluster of leaves at the end of very delicate shoots) the short-shoot chains cannot grow to any length or

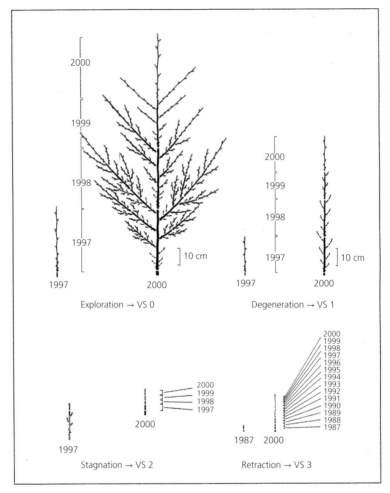

Figure 7.7 Growth stages of hardwood trees: on the left the state of the leader at the beginning of its development, on the right the development after 3 years (VS = vitality-class stages).

any age in the upper crown area exposed to wind and rain events. At this time, secondary factors determine the exact time of dieback. Typical claws are formed in the crown periphery, because the short-shoot chains, as they get longer, stretch toward the light.

Now it becomes clear why the different phases of growing are due to decreasing shoot lengths and, therefore, reflect a decreasing vitality.

Similar growth phases can be identified in every other investigated broad-leaved tree species, with only small species-dependant modifications. The following vitality class system is developed basing on these growth stages.

7.4.3 Vitality classes
Healthy vigorous trees of "vitality class 0" (Figure 7.8) show treetop shoots in the exploration phase: both the main axes and part of the lateral twigs consist of long-shoots. For this reason, a regular net-like branching pattern is developed, which reaches deep into the

Figure 7.8 Vitality classes 0–3 of a broad-leaved tree based on branching pattern (view of the leafless and leafy state, according to the growth stages in Figure 7.7).

interior of the crown. The crowns are equally closed and domed, and do not show any greater gap unless a stronger intervention has occurred, such as pruning measures, because such a gap is closed quickly by the intensive ramification. In this manner, the newly conquered airspace is quickly filled by the harmonic branching pattern. In summer, a dense foliage arises without any greater gap.

Weakened trees of the "vitality class 1" show treetop shoots in the degeneration phase. Thus, spears/"fox tails" are formed, rising above the canopy. The leaves on these spears are dense and grow all around them (at the top of the lateral short-shoots or short-shoot chains). The crowns make a frazzled impression on the outside, and have a fastigiate appearance, because the airspace between the spears is not completely filled by leaves and twigs, and the crown has a spiky outline. Inside the crown, the branching pattern, and hence the foliage, is quite dense. In this vitality class, straight percurrent main axes of the treetop branches are still dominant, but the crowns no longer look as intact as in class 0 because of the spears shooting out of the canopy.

In obviously less vigorous trees of "vitality class 2", the treetop shoots begin to build short-shoots in the stagnation phase. The leafless state could be designated as the claw stage, because the short-shoot chains in the outside of the crowns grow longer, are pre-dominant, and stretch claw-like to the light. These short-shoot chains, growing too long, break off in summer in thunderstorms and heavy rains, and strew the forest floor in declining stands. Under normal circumstances, trees get rid of parts of their unimportant twigs in the inner and lower crown parts in this way.

However, if the treetop shoots themselves are declining, the self-pruning of twigs progresses into the outskirts of the crown, and the crowns become thin from the inside outwards. The cause for this occurrence is not premature leaf fall, but broken short-shoot chains, a lack of shoots, and dead buds and twigs. The branching pattern shows a bushy and lumpy accumulation in the periphery of the crown. This accumulation causes summer and winter bushy crown structures and greater gaps. The crown periphery still has hardly any straight percurrent branches.

In considerably damaged or declining trees of "vitality class 3", the crowns finally fall apart by the breaking off of larger branches and the dieback of whole crown parts. The tree seems to consist only of more or less surplus sub-crowns, dispersed randomly in the airspace and forming whip-like structures. The treetop is often dying back or is already dead, because the treetop shoots grew in the retraction phase.

7.4.4 Vitality and tree life expectancy

Discussions abound as to whether this approach allows predictions regarding road safety, the future development, or even the life expectancy of the tree in question, based on its vitality. It is important to remember that this approach, being based on the branching structure, assesses the growth potential. It is safe to assume that a tree with a better growth potential will be better able to tolerate and compensate negative impacts (diseases, damage, deficiencies, pollution, etc.) than a tree that is less vital. This has been proven, for example, for compartmentalization in the case of fungal infestation or injury, for reactions to trimming (see Chapter 15) and for several diseases. Vigorous trees also have more capacity for photosynthesis (Rust & Roloff, 2002), are better able to buffer the stress due to their faster growth, and therefore have better expectations for the future.

However this does not allow for a prognosis regarding the remaining lifespan, just as you cannot tell how long a person is going to live based on how they feel (or look) today.

Figure 7.9 Vitality class 3 in *Tilia cordata* with dead leaders and dieback of the crown top.

Too many factors have an influence on life expectancy. Just as for people, it is impossible to predict how all of these factors will combine and develop over the years.

This also means that, if a tree is classified as vitality class 2 (trees with significantly reduced vitality – Figures 7.7 and 7.8) as a result of a vitality assessment, then that tree is not necessarily dangerous or needs to be cut down because it has no future! A tree with vitality class 2 can still be safe and can serve its purpose in the city for decades (e.g., if age is the reason for the poor vitality). The prognosis gets more critical when the tree reaches vitality class 2.5 or even 3 (Figure 7.9) – that is, when main axes of the crown start dying back. These particular warning signals are an indication of fundamental problems in the tree. In situations where road safety is a concern, it is usually necessary to at least (repeatedly) remove the deadwood.

After the previous explanations, it should be self-evident that such a vitality class system, based on criteria of the branching structures solely, can result from a long-term chronic diminution of vigor. This is the reason why the method is suitable to identify stress and tree decline in hardwoods. Therefore, modifications of the branching structure are due to such long-term negative factors affecting tree growth as root damage, soil compaction, chronic drought stress, air pollution, and so on.

This system which is based on crown structures may also be used successful in aerial photographs, thus making it possible to cover a larger forest or city area in a shorter time period (Figure 7.10).

Figure 7.10 Vitality classes 0–3 (from left to right) in aerial photographs (according to Figure 7.8).

7.5 Conclusions

The method was tested and confirmed in a variety of investigations (Stribley, 1993; VDI, 1993; Innes, 1994; Dobbertin, 2005; Dieler & Pretzsch, 2013). Recent research in many other hardwood species has shown the possibility of assessing the tree vitality in other areas of the temperate zone in the same way as reported here. In these hardwoods, only very few species-dependent modifications must be taken into account. The method is now successfully in tests for (sub)tropical trees (Davila, 2014: *Acacia* spec, *Ulmus* spec.; Marthalestari, 2014: *Acacia auriculiformis, Agatis* spec., *Ficus benyamina, Samanea saman, Swietenia* spec., *Tectonia grandis, et al.*).

References

Büsgen, M. and Münch, E. (1929). *The structure and life of forest trees*. Chapman & Hall, London.

Davila Siles, A.M. (2014). *Identification of Management Needs in Street Trees: a Pilot Study in El Prado of the City of La Paz, Bolivia*. Master Thesis TU Dresden, Dept. of Forestry.

Dieler, J. and Pretzsch, H. (2013). Morphological plasticity of European beech (Fagus sylvatica L.) in pure and mixed-species stands. *Forest Ecology and Management* **295**, 97–108.

Dobbertin, M. (2005). Tree growth as indicator of tree vitality and of tree reaction to environmental stress. *European Journal of Forest Research* **124**, 319–324.

Hallé, F., Oldeman, R.A.A. and Tomlinson, P.B. (1978). Tropical trees and forests. Springer, Berlin/Heidelberg/New York.

Innes, J.L. (1994). The occurence of flowering and fruiting on individual trees over 3 years and their effect on subsequent crown condition. *Trees* **8**, 139–150.

Lonsdale, D. (Hrsg.). (2013). *Ancient and other veteran trees: further guidance on management*. Tree Council, London.

Marthalestari, M. (2014). *Visual Tree Assessment of Tropical Urban Forest – A case Study from Jakarta Urban Forests, Indonesia*. Master Thesis TU Dresden, Dept. of Forestry.

Oldeman, R.A.A. (2014). *Forests: Elements of Silvology*. Springer, New York.

Roloff, A. (1989). Tree vigour and branching pattern. *Journal of Forest Science* **5**, 206–216.

Roloff, A. (2004). *Trees – Phenomena of Adaptation and Optimization*. Landshut.

Rust, S. and Roloff, A. (2002). Reduced photosynthesis in old oak (Quercus robur): the impact of crown and hydraulic architecture. *Tree Physiology* **22**, 597–601.

Stribley, G.H. (1993). Studies on the health of beech trees in Surrey, England: Relationship between winter canopy assessment by Roloff's method and twig analysis. *Forestry* **66**, 1–26.

Thiebaut, B. (1988). Tree growth, morphology and architecture, the case of beech: *Fagus sylvatica L*. In: Comm. Eur. Communities (ed.): *Scientific basis of forest decline symptomatology*, pp 49–72. Brüssels.

UN-ECE (2010). *Manual IV – Visual Assessment of Crown Condition*. http://icp-forests.net/page/icp-forests-manual. [accessed 22.01.2015]

VDI (Assoc. German Engineers). (1993). *VDI Guidelines: Determination of vegetational injuries*. Duesseldorf.

CHAPTER 8

Body language of trees, tree diagnostics

Andreas Roloff

Technische Universität Dresden, Tharandt, Germany

8.1 Terms and definition

Those who are much engaged with trees will be strongly impressed by how much of a tree's state, of its inner life and of its past/history can be deduced from its exterior appearance and symptoms. While trees cannot smile to indicate happiness, and do not look sad when they are suffering, as we humans do, events leave their marks on them for much longer – sometimes for life. We therefore also use the term "body language" for trees (Visual Tree Assessment/VTA by Mattheck and Breloer, 1997; Harris *et al.*, 2004) to mean the interpretation of external symptoms to assess the tree's current state, internal defects and the effects of past events.

The term "tree diagnostics" indicates that, in addition to the actual symptoms, a lot of background information and tree biology knowledge are taken into consideration, just as in human medical diagnostics (Roloff, 2015). Knowing the body language and the diagnostic characteristics is essential for tree inventory, assessment and understanding, where the primary objective is an initial evaluation of a tree's state and potential risks or defects. Possible causes of these can be diagnosed without using technical equipment, based on a visual assessment of external symptoms alone (see Chapters 7 and 9 and ISA, 2015). Depending on the results of this inspection, it may then be decided that further examinations (with technical devices if appropriate) should be carried out (see Chapter 9).

8.2 Adaptation and optimization in trees

Trees cannot survive without adaptation. They cannot escape colds, droughts or flooding, human damage or urban site stress. They can either cope with these and other events and factors, or they have to die. No other group of organisms on earth is as dependent on optimization and adaptation as trees, whose progress through generations is extremely slow by comparison to herbaceous or low plants, or to animals.

A tree can only survive by making the use of existing resources in the best way possible, by distributing its leaves in the air space and developing roots into the ground as efficiently as possible and at a minimum effort. This is known as the *survival strategy of trees*. In this context, the word "strategy" does not have the same meaning as it does for people; trees do not plan or calculate things. Still, it is common usage in ecology to speak

Urban Tree Management: For the Sustainable Development of Green Cities, First Edition. Edited by Andreas Roloff.
© 2016 John Wiley & Sons, Ltd. Published 2016 by John Wiley & Sons, Ltd.

of a strategy as features which have been adopted over the course of evolution and which facilitate survival among competitors. Strategy is the total sum of all genetically fixated physiological, anatomical and morphological adaptations that serve to conquer and to defend a site by using its resources as efficiently as possible (see Chapter 16).

Also, the term "strategy" does not refer to the individual tree, but to the entire species or the life form tree in general. A strategy ensures long-term survival, because it ensures the survival both of the individual trees and of the entire species by reproduction.

Trees can be understood as masters of survival, and their strategies of adaptation and optimization are admirable (see Chapter 16). If trees are seen this way, it may awaken an interest in imitating these phenomena, with a view to improving our daily life and our social existence and making it easier. Trees can, indeed, teach us a lot in this respect. They can serve as examples, especially for technical constructions and processes – an aspect which has recently been worked on by the science of bionics. This branch of science, however, which is still very young, has not been paying much attention to trees as yet.

Particularly impressive are the various mechanisms of self-repair and self-regeneration, the biological self-optimization which is taking place in trees at all times. However, we have to interpret these symptoms of self-repair on the basis of tree biology knowledge.

8.3 Examples and explanation: branches, trunk/bark, roots

We now will look into details of tree body language symptoms in the order of branches, trunk/bark and root/root collar.

8.3.1 Branch-shedding collar

A branch-shedding collar is a phenomenon which announces the imminent death of a branch by a collar-like diameter jump at its base (Figure 8.1). If the leaves of a certain branch cease to produce enough assimilates (e.g., because of shading or damage) it

Figure 8.1 Branch-shedding collars.

eventually becomes a burden to the tree, rather than being useful. In this case, it begins to die. The collar at the base of the branch prepares the healing process of the wound left by the death and breaking-off of the branch. Pruning methods which are based on tree biology take this collar into consideration, and it is crucial for it to remain intact (see Chapter 11). Branch-shedding collars develop only in some tree species, which is a pity, because many branches die without a collar.

8.3.2 Hazard beams

If a branch keeps sinking, it exceeds the tolerances of the wood mechanics at some point. The branch subsequently splits along the middle (through the pith), a dramatic symptom of wood failure (see Figure 8.2: branch in the background). This process starts with fine bark cracks, at which point the branch is then called a hazard beam. If it grows a few meters above a street or a path, it has to be removed immediately for safety reasons. The branch is about to break off and might fall out of the crown. This is a particularly impressive example of the body language of trees, as it indicates processes in the trunk and branches which would otherwise be invisible. For this reason, the bark of a tree is a very important symptom for inner trunk defects because, in it, the trained eye can recognize symptoms of stunting and stretching of the sides of the trunk and the branches.

Figure 8.2 Hazard beam.

Figure 8.3 Bottle butt.

8.3.3 Bottle butts

A bottle-shaped bulge at the butt end usually indicates that the interior of the trunk is rotting or hollow in this section (Figure 8.3). Wood rot diminishes conductivity and mechanical stability, which the tree tries to compensate for with excessive diameter growth. This leads to a swollen stem with bottle shape, often accompanied by bark fissures. Sometimes, the wood rot is already visible, even from the outside, but usually diagnostic devices or methods have to be used to achieve clarity. After all, the affected trunk section is usually of great importance for the mechanical safety of the tree.

8.3.4 Forked trees

So-called forked trees result if two leader shoots compete for the top position, and if this competition remains undecided. If the terminal bud, or the original leader of a tree, is lost or damaged, it has to be replaced by lateral buds/branches in so-called reiterations (see Chapter 7). Forks are a very important factor for (the assessment of) a tree's stability, because they can cause the tree to break apart. The steeper the angle at which the two stems are connected, the bigger the risk.

As they increase in diameter, the stems are pushing each other out of the way and, if the bark at the base coalesces, this can lead to wood rot. Depending on the shape of the angle between the stems, a distinction is made between V-forks and U-forks. U-forks are usually given a better prognosis regarding the stability of the tree. Figure 8.4 shows a V-fork, with a crack and included bark at its base.

Figure 8.4 V-forked tree.

8.3.5 Nose-like ribs on forked trees

A nose-like rib is a bulge which occurs in the trunk below a steep, V-shaped fork with two equally strong stems. It indicates considerable biomechanical problems at the coalescence of the two stems. The tree is trying to connect the stems at the base, by locally increasing its diameter growth (Figure 8.5). This is not entirely successful. Tension fissures (on the inside) constantly increase the incentive to grow. The mechanical risk of breaking apart is great in forked trees of this type. The phenomenon of nose-like ribs is another excellent example of the body language of trees, which provides a lot of information about their inner life and their history. Nose-like ribs can also occur on both sides of the trunk, in which case they are called ear-like ribs.

8.3.6 Sunburn

Trees can suffer from sunburn. It occurs if an older tree with thin or dark bark is suddenly replanted from a shady to a sunny site, or if the neighboring trees which used to shade it are removed (see Chapters 6 and 12). The south-facing side of the trunk then becomes so hot that tissues (bark and cambium) can die. The bark subsequently tears or flakes off. For this reason, it is important to maintain the direction of the crown whenever an older tree with thin bark is replanted, because the south-facing side of the trunk is already adapted to the heat. The same phenomenon also can occur after strong pruning measures in the upper crown (the lower branches used to be shaded by it), or if trees on the edge of a stand are felled (e.g., because of road works).

Figure 8.5 Nose-like rib.

8.3.7 Stem crack

Stem cracks can be caused by some very different factors, but the underlying cause is usually that tensions within the trunk are becoming visible. The immensity of the forces which are at work in the old trunk of a living tree often becomes recognizable after the tree is felled or, at the latest, when the wood is drying. The trunk then splits open or tears. As long as the tree, especially its mantle, is still intact, these inner forces contribute considerably to the mechanical stability of the tree. However, if defects occur or if the tree is subject to too much pressure, the tolerance of the optimized trunk is exceeded and it splits. This can also be caused by extremely fast growth, or by extreme differences in temperature in the trunk. Fissures are, therefore, an important warning sign in the assessment of a tree's safety.

8.3.8 Longitudinal splitting

If a trunk begins to lean over (due to a storm event or because the tension roots have been cut, etc.), this causes pressure on the side of the inclination and tension on the opposite side. The opposed forces can become so strong that the tree splits along the middle/pith. This is a dramatic warning sign indicating that the tree's stability is leaning over. Figure 8.6 shows a tree which has become dangerously inclined, even visible without the fissure. In other cases, the inclination is not quite so obvious, but the fissure symptom is still an important warning in the body language of the tree. It has great importance in the assessment of a tree's stability.

8.3.9 Knobs and nodules

Knobs and nodules can be caused by very different factors. Figure 8.7 shows an impressive grafting mark at the trunk base, where shoots from a purple-leaved beech tree were grafted onto a base of green-leaved beech when the tree was still young. To this day, the

Figure 8.6 Longitudinal splitting.

Figure 8.7 Knobs caused by grafting marks.

two different tree parts have not coalesced harmoniously. This has led to increased cell divisions, which have caused irregular growth in this area. Knobs can also develop for biomechanical reasons, in which case they are compensatory growths because of a defect on the inside of the trunk (wood rot, cavities etc.). Certain microorganisms (fungi, bacteria etc.) can also cause tumor-like knobs. They are usually benign, in that they only increase the growth in a limited and self-contained area.

8.3.10 Bark stripes on ribs

This phenomenon also belongs to the body language of trees. The bark on a rib tears open in stripes, or stripes of lighter, younger bark come to the surface (Figure 8.8). This indicates a strong diameter increase in this area. One reason may be that the tree is trying to improve the mechanical connection between a branch and the trunk, if the distribution of the branch's weight or its integration is problematic. In other cases, the diameter increase is due to particularly efficient assimilate production in the respective branch – that is, the branch is feeding an unusually high amount of sugar solution into the trunk. A look into the crown reveals which one of the two possible causes is the case. In the first case, the branch may have to be relieved of some of its weight by pruning, or it might even have to be removed – which would be completely wrong in the second case. If the tree is inclined, bark stripes can also occur on the upper side of the trunk, by tension wood in broad-leaved species.

Figure 8.8 Bark stripes.

8.3.11 Supply shadow

Branches which no longer achieve any remarkable photosynthetic gains due to long-term strong shading, or to leaf loss, no longer feed a lot of assimilates and hormones into the trunk. This can cause the diameter growth at the base of the branch to stay behind the surrounding trunk area (Figure 8.9). Eventually, a hollow rib can be formed. Since the direction of transport of the sugar solution is primarily down the trunk towards the root, this "supply shadow" only becomes noticeable below the branch base.

8.3.12 Elephant's foot

Large stones, rocks or curbs directly beside a tree's root collar can be used by the tree for mechanical anchoring. As a consequence, the tree often develops an "elephant's foot", which optimizes the mechanical forces in this part of the trunk by including the rock/kerb. The roots on the opposite side of the root collar have to be reinforced accordingly, because no rooting is possible in the direction of the rock/street. This can lead to the development of impressive growth anomalies at the root collar (see Figure 8.10). Smaller stones grow in, and eventually disappear completely in the root collar. From a physiological point of view, it is important that stones can retain a film of moisture on their surface for a long time. The tree can use this as an additional source of water during drought periods.

Figure 8.9 Supply shadow.

Figure 8.10 Elephant's foot.

8.3.13 Hollow trunks

A hollow trunk could be thought to be a threat to stability in any tree. However, in-depth calculations, studies and investigations have shown that this is not the case. Any grass blade or bamboo cane can be used to clarify this immediately: even two-meter-high grass survives a storm, even though the culms are almost entirely hollow. Trees also try to achieve maximum stability with the lowest possible effort and investment.

An important criterion for tree safety calculation is the relation of the thickness of the intact wall (t) to the trunk's radius (R). In practice, trees with closed, or almost closed, cavities are extremely unlikely to fail as the result of bending fractures (Figure 8.11). Where the thickness of the sound residual wall is less than 20% of the stem radius, failure results from the cross-sectional flattening. To come to an exact limit value between 10% and 30%, the crown size, tree species, exposure to wind, height, and so on, have to be taken into account.

8.3.14 Crown/root relationship

The crown and root of a tree do not grow independently, but have a certain relationship. For instance, the crown only keeps growing for as long as it can be supplied by the root system. At the same time, the root only keeps growing for as long as it can be supplied with assimilates from the leaf area. It has been shown that both parts of the tree aim for an equilibrium. The individual ratio depends on the species, the site, and on other environmental factors. If the equilibrium is disturbed (e.g., by pruning of the crown or

Figure 8.11 Hollow trunk.

by root loss due to construction work), growth of the other part of the tree has to stagnate (with a delay caused by stored assimilates). It cannot continue to grow normally before the equilibrium has been renewed. Lammas shoots indicate that the roots are growing faster than the crown.

8.3.15 Root symphysis
Root symphyses (anastomosis), which are normally invisible, become recognizable when the root collar of a felled tree continues to increase in diameter, or when the cut surface is being occluded. It can then be deduced that the root collar is supplied with sugar solution (which is no longer provided by its own crown) by a living neighboring tree. Studies in different tree species have shown more than 50% of the trees can be connected by symphyses, and that they communicate and exchange substances via their roots (www: "wood-wide web"). Symphysis also increases the stability of a stand or an avenue. At the same time, however, it allows pests (e.g., fungi) to spread faster. With many diseases, this can be disastrous.

8.3.16 Tension roots on slopes
The most important roots for the stability of a tree which is growing on a slope are the surface roots which are oriented up the slope. The strongly developed root collar usually identifies them as tension roots (see Figure 8.12). These roots are more important for the stability of the tree than the lateral roots and the roots growing down the slope, due to the fact that wood is optimized in tensile strength. Trees growing on slopes or riverbanks are, therefore, held in place by the traction of the tension roots rather than

Figure 8.12 Tension root on a slope.

pressure from the downhill-pointing roots. Tension roots are therefore usually particularly well-developed, and they have to be intact in order for the tree to stand securely. Hence, damage (e.g., injuries or wood rot) to the tension roots must be considered particularly dangerous to the stability of the tree. It is the same with roots orientated to the main wind direction.

If the tension roots are cut in the process of construction work, or damaged by fungi, the tree becomes extremely susceptible to windfall. Older trees are unable to compensate for the loss of their tension roots. The problem occurs frequently when streets are extended, or when cycle paths are built. The tension roots of the trees are cut in the process, and the trees loose their stability (if the main direction of the wind is from the road side). If the same amount of roots is removed on the other side, the risk to the stability of the trees is much lower.

8.3.17 Covered root collars

The only explanation for a missing root collar in an older tree (except palm trees) is that it has been covered up. This can happen naturally, due to flooding along rivers or due to landslides. In cities, construction work is often the cause. Most species are very sensitive to such a change in their environment, because they are unable to penetrate the higher top soil with new roots, and because the roots which used to be close to the surface are now in the more oxygen-deficient, lower layers of the soil (see Chapter 4). They, and their mycorrhizae, are also compressed and often die. In combination with wood rot, this

Figure 8.13 Root collar strangling.

often eventually leads to the death of the entire tree. Species which naturally grow in constantly changing habitats (e.g., in riparian forests) deal much better with these changes, whereas others are particularly sensitive.

8.3.18 Root collar strangling
If the root collar is strangled (as in Figure 8.13, because the ground was sealed) the roots do not receive enough sugar solution from the crown. This means that the production of fine roots decreases which, in turn, leads to a water deficit. The result is a vicious circle which the tree can no longer escape. It is very likely to die (this is probably intended, in the example in the photo in Figure 8.13). The swelling which is visible above the strangled section is often interpreted as a congestion of assimilates. The actual explanation is increased growth in this area; the sugar solution from the crown cannot continue its way down to the root.

8.3.19 Sealing of the root area
If the root area is sealed (Figure 8.13), the supply of water and nutrients to the tree is reduced. Subsequently, the crown can no longer be provided with a sufficient amount of water, and the tree slowly begins to die. It is decisive for the prognosis whether the seal has always been there (in which case, the tree would have found water in the course of its development), or whether the ground has been sealed only recently. In the latter case, and if the seal is strong, it leads to a loss of large parts of the root system. Older trees then

Figure 8.14 Inner roots.

react by dying. Some tree species deal better with such a change than others, depending on their natural habitat. The older the tree is at the time of sealing, the more critical its reaction will be.

8.3.20 Inner roots

In very old trees (which are always hollow), it can often be observed that inner roots have developed within the hollow trunk (Figure 8.14). They grow from a point up to two meters above ground, on the inner mantle of the trunk, and grow downwards in the semi-darkness of the cavity. If they eventually manage to reach the ground, they become secondary roots of the trunk mantle, contributing to the stabilization of the torso and supplying it with extra water and nutrients. For this reason, secondary roots actually have a very important function in the complicated metabolic processes of an ancient tree. Inner roots are much more common in *Tilia* and *Ficus* trees than in other species. They are one of the reasons why these trees can live so long.

8.3.21 Adventitious roots

If parts of the crown have been flooded, the branches can develop so-called adventitious roots. The faster and the more intense this reaction is in a species, the more efficiently it can deal with the flooding. Lack of oxygen is a great problem for the submerged branches and roots (see Chapter 4). The newly formed roots are, therefore, specially adapted to the

Figure 8.15 Adventitious roots.

changed conditions (due to a high percentage of air-conductive tissues), and can thus ensure survival. This method works particularly well in willows, which can form adventitious roots very quickly (Figure 8.15). Some other species are completely unable to form adventitious roots, which causes them to die rather quickly if submerged. They do not survive in natural riparian forests for very long, whereas willows often grow on the edge of a watercourse. Adventitious roots also develop if a tree is covered by soil, in which case similar rules apply.

8.4 Conclusions

Knowing the symptoms of tree body language is an important tool in monitoring, evaluating and diagnosing trees. However, we must not forget that this approach has its limits. Many symptoms may have a number of different causes, leading to very different consequences. Tree biology knowledge and experience are, therefore, indispensable in order to correctly understand, interpret, and evaluate the possible causes (see Chapters 2 to 16 and: Buesgen and Muench, 1929; Zimmermann and Brown, 1971; Kozlowski *et al.*, 1991; Shigo, 1994; Kramer and Boyer, 1995; Kozlowski and Pallardi 1997; Lambers *et al.*, 1998; Tyree and Zimmermann, 2002; Larcher, 2003; Kimmins, 2004; Schweingruber *et al.*, 2006; Taiz and Zeiger, 2006; Schweingruber, 2007; Bowes, 2010; Matyssek *et al.*, 2010).

References

Bowes, G.B. (2010). *Trees and Forests – A Colour Guide*. Manson Publ., London.

Buesgen, M. and Muench, E. (1929). *The structure and life of forest trees*. Chapman & Hall, London.

Harris, R.W., Clark, J.R. and Matheny, N.P. (2004). *Arboriculture*, 4th edition. Pearson Education, New Jersey.

ISA (International Society of Arboriculture) (2015). *Tree risk assessment form. www.isa-arbor.com [last accessed 01.02.2015]*

Kimmins, J.P. (2004). *Forest Ecology*, 3rd edition. Prentice Hall, New Jersey.

Kozlowski, T.T. and Pallardi, G. (1997). *Physiology of Woody Plants*. Academic Press, San Diego, CA.

Kozlowski, T.T., Kramer, P.J. and Pallardy, S.G. (1991). *The Physiological Ecology of Woody Plants*. Academic Press, San Diego, CA.

Kramer, P.J., Boyer, J.S. (1995). *Water Relations of Plants and Soils*. Academic Press, San Diego, CA.

Lambers, H., Chapin, F.S. and Ponis, T.L. (1998). *Plant Physiological Ecology*. Springer Verlag, New York.

Larcher, W. (2003). *Physiological Plant Ecology*, 4th edition. Springer, Berlin/New York.

Mattheck, C. and Breloer, H. (1997). *The Body Language of Trees – A Handbook for Failure Analysis*, 3rd edition. Rombach/Forschungszentrum, Karlsruhe.

Matyssek, R., Fromm, J., Rennenberg, H. and Roloff, A. (2010). *Biologie der Bäume*. Ulmer, Stuttgart.

Roloff, A. (ed.). (2013). *Baumpflege*, 2nd edition. Ulmer, Stuttgart.

Roloff, A.(2015). *Handbuch Baumdiagnostik* – Baum-Körpersprache und -Beurteilung.

Schweingruber, F.H. (2007). *Wood structure and the Environment*. Springer Verlag, Berlin/Heidelberg/ New York.

Schweingruber, F.H., Börner, A. and Schulze, E.-D. (2006). *Atlas of Woody Plant Stems*. Springer Verlag, Berlin/Heidelberg/New York.

Shigo A. (1994). *Tree Anatomy*. Shigo & Tree Assoc., Durham.

Taiz, L. and Zeiger, E. (2006): *Plant Physiology*, 4th edition. Sinauer Assoc., Sunderland/USA.

Tyree, M.T. and Zimmermann, M.H. (2002): *Xylem Structure and the Ascent of Sap*, 2nd edition. Springer Verlag, Berlin/Heidelberg/New York.

Zimmermann, M.H. and Brown, C.L. (1971). *Trees – Structure and Function*. Springer Verlag, Berlin/ Heidelberg/New York.

CHAPTER 9

Tree inventory, risk assessment and management

Steffen Rust

HAWK Hochschule für angewandte Wissenschaft und Kunst, Fakultät Ressourcenmanagement, Göttingen, Germany

9.1 Introduction

Trees cannot be maintained free of risks. As trees become older and grow larger, the benefits they provide to society increase, but so does the likelihood of failure. The National Tree Safety Group, an organization based in the United Kingdom, has identified principles to balance tree risks and benefits (NTSG, 2011):

- Trees provide a wide variety of benefits to society.
- Trees are living organisms and naturally lose branches or fall.
- The risk to human safety is extremely low.
- Tree owners have a legal duty of care.
- Tree owners should take a balanced and proportionate approach to tree safety management.

This is very similar to statements in the tree inspection guidelines of the FLL (Landscape Development and Landscaping Research Society) in Germany (FLL, 2004, 2010), and it is also emphasized in guidance on tree risk assessment by the ISA in North America.

Few studies have analyzed the risk to human safety quantitatively. A report from the UK estimates the risk of being killed by a tree as less than one in ten million per year. Nevertheless, in many countries, health and safety legislation requires tree owners to have a suitable and sufficient risk assessment, and to apply measures that are reasonable and practicable.

Tree risk management is the application of policies, procedures and practices used to identify, evaluate, mitigate, monitor and communicate tree risk (Dunster *et al.*, 2013). This chapter covers tree inventories and risk assessment, while Chapters 10 and 11 treat mitigation.

Much progress has been made in recent years to formalize the way that trees are assessed. Two important institutions from the arboricultural industry – ISA in North America and FLL in Germany – have produced guidance on the different levels of tree assessment, risk categorization (ISA), and appropriate time frames (FLL). This provides a basis to define the scope of work for arborists and clients.

Organized risk management not only helps to identify and control the risk, and to comply with relevant legal and regulatory requirements, but also to improve stakeholder trust. Often, local residents will oppose the decision to remove a tree that is considered a risk. Transparent and well-documented procedures, together with early participation of

Urban Tree Management: For the Sustainable Development of Green Cities, First Edition. Edited by Andreas Roloff.
© 2016 John Wiley & Sons, Ltd. Published 2016 by John Wiley & Sons, Ltd.

stakeholders, will help to reduce conflicts. Human safety is only one part of the management of green spaces. It is therefore important to recognize that risk management can be undertaken only by understanding the trees and their value to people in the context within which they grow.

This chapter reviews the basics of tree inventories, explores recent concepts of tree risk assessment, and ends with an overview of modern – mostly non-destructive – advanced tree assessment methods.

9.2 Tree inventory

Quantitative knowledge of the resource should be the foundation of tree management. A tree inventory provides data such as number, location, size, and species to support negotiations on budgets for tree care, allocate resources and organize tree management. It provides a framework for the documentation of tree assessment, keeping records of tree inspections and maintenance. If litigation involves a tree, this can provide evidence that the municipality has not been negligent, thus reducing liability for damage or injury.

Inventory data can be used to valuate the many environmental and aesthetic benefits provided by trees, including storm water control or increased property values, through a recognized valuation system (McPherson, 1992, 2007). Examples are i-Tree of the USDA Forest service, CAVAT in the UK, or the "Koch Method" in Germany. Expenditures for trees can then be balanced with a monetary value. When management and maintenance budgets are allocated in proportion to the asset, this can be an opportunity to increase the tree budget.

It may not always be necessary to implement a full inventory. A sophisticated sampling strategy can provide most of the data for valuation, budgeting and planning. Urban tree inventories can be based on data from satellites, airplanes, on-the-ground scanning or digital photography and field surveys. At present, the reliability of the data obtained from all methods except field surveys is limited (Nielsen *et al.*, 2014).

9.2.1 Inventory parameters

Most inventories contain basic descriptive information about location, species and size of the trees. Diameter at breast height correlates with many other tree traits, including height, crown diameter or age, and can thus be used as a proxy for other, harder to estimate parameters. Tree height is useful for work planning, for example, because the size of equipment like aerial lifts can be adequately chosen. Information on conflicts (e.g., with utilities) and growing space is also valuable. Often, the frequency of tree risk assessment depends on parameters such as occupancy, targets and age of the tree, so these need to be included.

The results of tree inspections, including the name of the tree assessor, the date of inspection, recommended actions and their timing for risk mitigation, are frequently further parameters of inventories. Any tree work, especially when intended to mitigate risks, should be documented.

9.2.2 Technology

Tree data can be collected by in-house staff, contractors, or volunteers. Each approach has its own advantages.

Several recent projects have demonstrated how members of the general public can, with adequate training, help to map and monitor trees. Using collaborative online mapping tools, citizens can collect large amounts of data, as in the Ancient Tree Hunt in the UK. Although volunteers are inexpensive and can provide a general overview of the urban trees, they should not be relied on for tree risk assessment. Where certification schemes for tree assessors exist, such as those of ISA in North America or FLL in Germany, municipalities might require certificates or equivalent qualification to ensure minimum standards for contractors and in-house staff.

Often, in-house staff does inspect the same trees for many years. This is a major advantage, because processes like decay and decline can be very slow, and their significance can be overestimated by consultants assessing a specific tree only once in its lifetime. Often, though, municipalities do not have sufficient staff, or a low number of trees does not justify the employment of specialized staff, so that contractors are employed.

When out-sourcing the tree inventory and risk assessment to consultants is considered, it should be taken into account that the overheads of supervision and quality control might outweigh the often low per-tree costs of contractors.

There have been rapid changes in the hardware and software markets. Dedicated, single-purpose proprietary software, add-ons to existing platforms, cloud-based services, or software as a service, might differ in their sustainability.

The choice of technology ranges from pencil and paper to dedicated computer hardware and software. While records on paper may satisfy legal requirements and can be appropriate for small communities with only a few trees, further data analysis is hardly possible. The transfer of data to spreadsheets or databases is prone to errors, but it can be a cheap option, especially when free software is used.

The choice of dedicated, proprietary software solutions is large. While some companies rely on rugged outdoor PCs, others use tablet computers or mobile phones. Apart from practical considerations of the people intending to use the hard- and software during the survey, the accessibility of the data by a wide range of users should be ensured.

In most cases, individual trees must be identifiable for inspection and management. This can be achieved with labels on each tree, or with accurate information on their position. Labels should not harm the tree. They should be attached at least 2.3 meters above ground in order to protect them from vandalism.

Internationally, several institutions offer guidance for tree inventories: ISA has formulated Best Management Practices, in Austria a national standard has been published, in Germany there are guidelines by the FLL and, in Sweden, Östberg *et al.* (2013) published standards based on their scientific studies.

9.3 Tree risk assessment

This topic will be outlined on the basis of the example of two recent and widely adopted, but differing approaches. One has been developed by the Landscape Development and Landscaping Research Society (FLL) in Germany (FLL, 2004, 2010, 2013), and the other by the International Society of Arboriculture (Dunster *et al.*, 2013).

9.3.1 Terms and concepts
Terms and concepts are used here as in the ISA Tree Risk Assessment Manual (Dunster *et al.*, 2013). Tree risk assessment as part of risk management (Figure 9.1) is the systematic process of identifying, analyzing and evaluating tree risk. Risk is the combination of the likelihood of a conflict or tree failure occurring and affecting a target, with the severity of

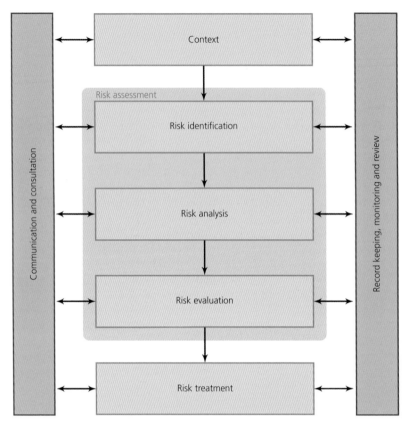

Figure 9.1 Risk assessment in the risk management process (ISO 31010). (Adapted from Dunster et al. 2013.)

any associated consequences, such as personal injury, property damage or disruption of activities. Tree risk evaluation is the process of comparing the assessed risk against given criteria in order to determine the significance of that risk.

While most of this chapter focuses on risks due to structural failures, conflicts between trees and other infrastructure or functions need to be assessed, as well. Trees may obstruct traffic signs, shade solar panels, produce annoying fruits or allergenic pollen, or lift pavements.

Most tree risk assessment is qualitative rather than quantitative, because numerical probabilities of failure are rarely available for trees.

The intensity of assessment is defined as different levels: ISA's *Best management Practices: Tree Risk Assessment* defines three levels (limited visual, basic, and advanced) while, in Germany, FLL defines two levels only, which correspond to ISA's "basic" and "advanced". A limited visual assessment is intended to quickly collect information on large populations of trees, looking for obvious defects while walking, driving, or even flying past the trees. While frequently this level does not fulfill the requirements for risk assessment, it can be very useful after storms, along power lines, or in forests, where legal requirements are often lower than in urban areas.

A basic assessment is a detailed visual inspection of the entire tree, both from a distance and close up, staying on the ground, optionally using simple tools. If the basic assessment is not sufficient to evaluate the risk of failure of a tree, advanced assessments can provide

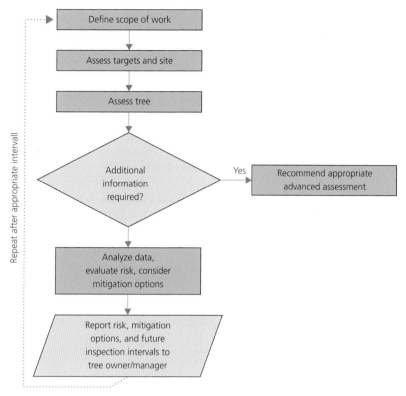

Figure 9.2 Flowchart of basic level of assessment. (Dunster et al. 2013. Reproduced with permission of the International Society of Arboriculture.)

specific information about the tree, targets or the site (Figure 9.2). Generally, only a small fraction of all inspected trees – often less than one in a thousand trees – should require advanced assessments, which are expensive because they are time-consuming and need special equipment and training.

9.3.2 Visual assessment
Visual assessment is the basis for all levels of assessments. A basic assessment will take approximately 2–20 minutes to inspect the tree, its site and possible targets.

9.3.2.1 Site assessment
The site, and changes to the site, can have a major impact on tree stability. They affect wind load, anchorage, and vitality of trees. The history of tree failures and of changes in land use can be very informative (Figure 9.3).

The local forces that wind generates in the trunk exceed by far those caused by the weight of the tree. Wind is the most common cause for tree failures (Metzger, 1893; Jacobs, 1936; Mergen, 1954). The size and structure of its crown, the surrounding terrain, and shelter provided by neighboring trees or buildings determine the wind load a tree is exposed to (Metzger, 1893; Tirén, 1928; Niklas, 1992; Vogel, 1994; Cullen, 2005). On the largest scale, maps of wind zones provide estimates of wind speed that might be expected in a specific region. Rough terrain, such as in inner cities, will reduce average wind speed, but can increase gust wind speed, as do funneling effects between buildings, slopes and ridges.

Figure 9.3 Sites can change dramatically. Knowing the history of a tree will help to assess it. The apparently favorable site of this maple (right) is heavily compacted, and the soil level has been raised.

However, trees that have lived most of their lives in an exposed condition are usually well adapted to wind. When neighboring buildings or trees are removed, previously sheltered trees can be more likely to fail for some years, until they have adapted to the increased wind load.

The roots of many tree species do not survive for long in waterlogged soil, and suddenly raising the water table will kill parts of the root system, leaving the tree unstable for years (see Chapter 4). A permanently high water table in irrigated landscapes can result in shallow rooting. High soil water content, as after heavy rains or floods, can reduce the shear strength of the soil and, thus, the anchorage of trees. Retreating ground water in the vicinity of wells or construction sites can cause an increase in dead branches, due to drought stress.

Soil compaction caused by vehicles or pedestrians will increase the concentration of carbon dioxide and reduce the concentration of oxygen in the soil. This can result in less fine roots, reduced vitality, and an increase in deadwood (Weltecke and Gaertig, 2012; and see Chapter 4).

Excavation, trenching, lowering the existing soil grades, and driving heavy machinery across tree roots, can damage roots and reduce tree stability. Although the immediate reduction in anchorage strength can be surprisingly low, decay will set in and might eventually – often several years after the causal event – result in root system failure.

Raising soil grades can similarly damage the roots of some species, even where only a few centimeters of soil block aeration. Although a new root system may develop close to the new surface, these roots may not be sufficient to replace the anchorage provided by the old structural roots.

The bark of young – and in some species even mature – trees can be severely damaged when exposed to direct sunlight, resulting in decay and structural weakening of the stem (see Chapter 6). The removal of neighboring trees and buildings that are shading the stem can cause the trees to fail some years later.

An assessment of the site and recent, ongoing or planned changes to it will help to assess the trees and the likelihood of failure of these trees.

Figure 9.4 The target zone of this veteran oak is frequently occupied by picnicking kindergarten groups.

9.3.2.2 Target assessment

People who could be injured, property that might be damaged, or activities that could be disrupted by failing trees, are considered "targets" in tree risk assessment, which may be carried out either if required by law, or if the potential damage to the target justifies the cost. While people will be the most important targets (Figure 9.4), the disruption of utilities such as electric power lines or important access roads by failing trees can have tremendous consequences for many people.

The area relevant to identify targets is the target zone, which is the area in which the tree or tree parts are likely to fall when they fail. The size and shape of the target zone depend on the tree and the terrain. It can be circular for a tree in level terrain that is not leaning, or asymmetrical on a slope or for a leaning tree. Because parts of falling trees might scatter farther than the height of the tree, a radius of up to 1.5 times the tree height is sometimes chosen. If the falling tree could impact a neighboring tree, the radius might even be larger. The target zone of a branch is the area where it might land, depending on its height of attachment and aerodynamic properties.

Targets in the target zone can be static (e.g., a building or a highway), movable (e.g., picnic tables) or mobile (e.g., pedestrians or cars). Movable targets might be relocated out of the target zone to reduce the level of risk.

The likelihood of a target being impacted depends on the time spent in the target zone – that is, on the occupancy rate. The occupancy rate can be classified as constant,

Figure 9.5 A fence prevents people from entering the target zone of this *Pterocarya fraxinifolia* with broken branches and failed cables. This has reduced the occupancy rate from frequent to rare.

frequent, occasional, or rare. Rare occupancy means that the target zone is not commonly used by people, while occasional indicates infrequent or irregular use. A target zone is occupied frequently when there are people for a large part of the day or week (Figure 9.5).

Four categories might be used to categorize the likelihood of a target being impacted:

- Very low: the likelihood that the failed tree or part impact the target is remote. This can be the case if the target zone is rarely occupied, or some structures provide protection against a target being impacted in an occasionally occupied area.
- Low: it is not likely that the failed tree or part will impact the target. This might be the case in an occasionally occupied area that is fully exposed to the tree, or a frequently used area that is partially protected.
- Medium: the failed tree or part is as likely to impact the target as not. This might be the case in a frequently used area exposed to one side of the tree, or a partially protected, constantly occupied area.
- High: the failed tree or part will most likely impact the target. This is the case for constant occupation and full exposure to the tree, or for a frequently occupied road or walkway next to the tree.

The consequences of failure are a function of the value of the target and the potential injury, damage, or disruption that could be caused by the impact of the failure. Four categories of consequences are described in the ISA qualitative tree risk assessment:

- Severe.
- Significant.
- Minor.
- Negligible.

Examples of severe consequences are fatalities, disruption of a primary power line, or blocking of an emergency access. Minor consequences include damage to low-value targets, while a small branch striking a fence can be considered a negligible consequence.

Ranking of targets according to value and importance can be used to prioritize targets for assessment and mitigation. An example is the guidance of FLL, which recommends frequencies of risk assessment based, among other criteria, on targets and occupancy.

9.3.2.3 Frequency of assessments and assessment time frame

The frequency of assessments of a tree depends on its state, its site and possible targets. The FLL guidelines widely adopted in Germany suggest intervals of one, two, and three years for trees planted more than approximately 15 years ago. These guidelines do not use a risk assessment, as in the ANSI standard. Younger trees do not need specific inspections, but are assessed along with regular tree maintenance. Maturing trees younger than 50 to 80 years are assessed every three years, when they have no or minor defects and are located at a site where a reasonable user should expect low safety standards (e.g., infrequently used paths or parks). If a tree is located at a site where users can expect higher safety standards, the suggested frequency is two years. Mature trees with no or minor defects are inspected annually, unless they grow at a site where reasonable users expect low safety standards, when inspection frequency is biannual. For trees with more severe defects, that might become a hazard within the next two years, an annual inspection cycle is recommended.

Often, the inspection interval is used as the assessment time frame. This time frame should be specified, and it should be indicated that an earlier or additional inspection may be necessary after extreme weather events, or when the site has been changed considerably. Examples are potential loss of roots or shelter provided by neighboring structures.

9.3.2.4 Tree assessment

At any level of risk assessment, one tries to determine how likely it is that the tree, or a part of it, will fail within the assessment time frame. Because every tree (or parts of it) can fail under extreme circumstances, the risks relevant for risk assessment must be concrete and foreseeable in normal climate conditions.

Trees or their parts will fail if and where a load exceeds the load-bearing capacity. In most cases, neither load nor load-carrying capacity are known. Often, the load-carrying capacity of the tree and its parts are several times higher than loads caused by their own weight and wind storms. Tree assessment looks for defects or structural conditions that might reduce strength or anchorage, but trees can compensate for strain increasing around a defect with additional growth and tougher wood (see Chapter 8). This should be taken into account when evaluating the risk of failure, and it is often the first indicator of a hidden defect.

When trees break, very often their failure starts with the buckling of fibers on the compression side of the trunk or branches, which happens when wood is compressed beyond its elastic limit. The extent of compression is a function of the force normal to the axis, the lever arm, and the flexural stiffness of the cross-section.

The main force acting on trees is drag from wind although, in some regions, ice and snow can be equally destructive. Its size depends mainly on wind speed, the projected area normal to the wind, and crown structure. Wind speed increases with height above ground. It transfers momentum to the tree via pressure drag and skin friction. Leaf and crown reconfiguration can significantly reduce drag at high wind speeds.

Static loads are relatively constant over a period. Wind load is dynamic and is dominated by rapid changes in forces, so that damping and resonance can become relevant. Whether resonance contributes to failure in urban trees is yet unclear, however.

Thus, trees or parts of trees that expose significant surface to the wind, and are attached to the rest of the plant with a long lever arm, can be expected to experience high loads. Bends, cracks, old pruning wounds and so on can locally raise stress in the structure.

Flexural stiffness is the product of the modulus of elasticity and the second moment of inertia. There is, however, no practical way to measure the modulus of elasticity for tree assessment. Static pulling tests can estimate flexural stiffness, and the second moment of inertia – which is a function of the geometry of the stem or branch – can be assessed with tomographs. Usually, however, the relative loss of cross-sectional strength due to a cavity is assessed by one or more estimates of residual wall thickness. Commonly, trees with a remaining wall thickness of one-third of the stem radius are considered unlikely to fail. There are, however, numerous limitations. Pertinent formulas assume one circular and central cavity in a circular stem of an unpruned, solitary tree.

Slender trees or branches could, in theory, have higher stresses at their base, if all else remains equal. Usually, though, slender trees have smaller crowns and, thus, wind load. They also have stronger wood and, thus, higher load-carrying capacity than their less slender counterparts. Most empirical evidence shows that slender trees and branches do not have a higher risk to fail.

The uncertainty of measurements of tree height, stem diameter and branch length is high, and that of the ratios of uncertain measurements like slenderness or residual wall thickness is even more so.

A major part of tree assessment is to identify defects that might lower the load-carrying capacity of the tree. In the present context, defects include injuries, decay, growth patterns, and other conditions reducing the structural strength of the tree. In the tree risk assessment process proposed by ISA, the impact of these defects on the likelihood of failure of the tree within the time frame of the assessment has to be classified.

The categories used in this system are: improbable (the tree or branch is not likely to fail during normal weather conditions); possible (failure could occur, but is unlikely during normal weather conditions); probable (failure is expected during normal weather conditions); and imminent (failure has started or is most likely to occur in the near future, even without significant wind or increased load). The factors to consider for a suitable time frame are outlined above.

9.3.2.5 Crown

In a systematic procedure, all parts of the tree are scrutinized. In the crown, dead, broken or hanging branches are of particular concern. Although, in some species, dead branches can remain mechanically stable for many years, their likelihood of failure will range from possible to imminent. Broken or hanging branches are probable or imminent to fail. As a rule, branches larger than 3–5 cm in diameter can cause injury and damage, although even smaller ones falling from great height can have severe consequences. Curved branches straightened by wind or their own weight can split along their neutral plane. This is termed a hazard beam (see Chapter 8). Often, such branches are too large to be pruned from the tree entirely without large wounds. To reduce the risk of failure, they can be reduced in length and their movement can be restricted with cables.

Co-dominant branches and stems are less well attached to another than is the case when one part is clearly dominant (i.e., is more than twice the diameter of the other). Sharp V-shaped junctions can include bark, weakening the attachment further, as do decay and cracks. In many species, co-dominant stems, and branches with included bark, have a likelihood of failure from possible to probable. If there is decay or there are cracks, the likelihood of failure is probable to imminent.

Phytotelmata between codominant stems do not increase the risk of decay or rupture due to ice expansion. Instead, they can provide important habitats.

Epicormic shoots, which especially form after topping or stem breakage, can be attached weakly, due to decay in the cut or broken end. Topped trees can develop extensive decay and cracks.

Tree support systems in the crown, like cables, should be inspected for defects, corrosion, correct installation and dimension, durability date, and abrasion. Support systems remain in the crown for many years while the tree is growing in height and weight. After some years, a once correctly installed support system might be too low within the crown, or not strong enough for the increased loads. Fibers lose their strength, due to UV radiation and wear. Therefore, many products have color coding schemes to identify installation date from the ground, and to decide whether the end of the warranty period, often 8–12 years, has been reached.

Cavities created by woodpeckers indicate decay of at least the size of the bird and its offspring. If this is large, compared to the branch it is located in, it constitutes a risk of failure. For further inspection and maintenance, national and international laws protecting endangered species and their habitats have to be respected.

Epiphytes like mistletoe can cause deformations and die-back of branches (see Chapter 6). Our recent experiments on *Populus* and *Tilia* indicate that deformations caused by *Viscum album* actually increase the strength of branches in one of the species, but not in the other. Epiphytes can impede the inspection of crotches. However, they provide habitat and food for animals, and should only be removed when a hazardous condition is likely to be concealed behind them. Epiphytes can increase the load of wind, ice, snow, and rain.

Fruit bodies of fungi causing wood decay can indicate an increased risk of failure in branches, depending on the species and type of decay it causes. Aerial inspection might be required to assess them (see Chapter 6).

Often, the vitality of trees is assessed on the basis of features of the crown. Although it is not directly related to risk, it contributes to decision-making for maintenance or replacement of the tree (see Chapter 7).

9.3.2.6 Stem

The stem should be assessed for any indication of reduced strength, including cracks, decay, fruiting bodies of fungi (see Chapter 6), or open cavities.

Collision of vehicles is a frequent cause of wounds in urban trees. Commonly, the damage is quite superficial. Often, features of the bark indicate locally reduced wood increment, which might be caused by underlying decay that killed the cambium. In such cases, sounding with a mallet allows the extent of the damage to be assessed. Activity of burrowing insects like ants or termites is an indication of decay or cavities in the stem, and might warrant a closer look.

In most countries, cavity treatment was stopped two or three decades ago. Therefore, signs of cavity cleaning and excavation or rods across open cavities will indicate a long-standing problem that was large enough to warrant surgery a long time ago. Some formulae for strength loss take open cavities into consideration. In these, an opening larger than one-third of the circumference is considered an indication of a risk of failure. The depth of open cavities can be probed with a bicycle spoke or similar tools.

9.3.2.7 Stem base, root collar, and site

Often, trees do not grow straight upright, due to competition with neighboring trees or shading by buildings. These trees are adapted to the lean, and will have an increased risk of failure only in special circumstances – for example, when the inclination of the stem is very high, and the soil is saturated with water after flooding or heavy rain. If, however, the lean results from partial failure during a storm, there is initially a high risk of failure, although this declines with time.

A lack of above-ground root flare can be the consequence of soil infill. Impervious soil on top of the original soil level will reduce the exchange of carbon dioxide and oxygen. Many tree species do not tolerate more than a few centimeters of infill, and their roots will consequently die (see Chapter 4), resulting in reduced anchorage strength. Evidence of soil excavation or compaction is a cue to inspect the stem base closer.

9.3.2.8 Clearance zone

Trees may not be allowed to grow into road clearance zones or to obstruct traffic signs or traffic lights. In Germany, for example, 4.5 m above roads, and 2.5 m above foot and bicycle paths, should be kept clear of branches and stems, as well as 0.5 m to either side. It is not always necessary, or possible, to remove branches or stems. In this case, traffic signs and beacons can warn and direct traffic. Also, visibility splays and sight lines must be kept clear; this includes trunks and branches of trees (see Chapter 11).

9.3.3 Advanced assessment

Detailed visual assessment is the starting-point of all further investigations. Done by an expert, this might already be sufficient to evaluate the tree. Crown inspection, either from an aerial lift, a ladder, or by climbing the tree, will help to assess features invisible from the ground, such as cracks, included bark, or decay on the upper side of branches.

This section focuses on the tree and does not cover advanced assessment of the site or targets. Most advanced methods are used either to measure the extent of internal decay, to assess root loss and its effects on anchorage, or to investigate pathogens.

9.3.3.1 Internal decay

Physical properties of the wood, the geometry of the stem and the decay, or the reaction of the tree when a load is applied, are used to assess the likelihood of failure due to internal decay (Figure 9.6).

Most often, the strength of a tree part is assessed on the basis of the distribution of sound wood (Wagener, 1963; Mattheck *et al.*, 1994; Kane and Ryan, 2003; Kane *et al.*, 2004; Bond, 2006) – that is, on the second moment of inertia. The size and position of decay, cracks, or included bark in the trunk or branches can be measured with a variety of tools. They vary in terms of damage to the tree, spatial resolution, and parameters measured. Often, these methods neglect the extent of decay along the stem axis, since they only determine a one- or two-dimensional image of the hollowed section. However, recent research has shown that the length of the decay column has a significant effect on the strength of tubular plant stems.

Drilling devices such as increment borer, drill bits or resistance-recording drills provide information about the wood along one line. While the remaining wall thickness is the only immediate quantitative result of increment cores, results from resistance-recording drills can correlate with wood density along the drilled path (Rinn *et al.*, 1996).

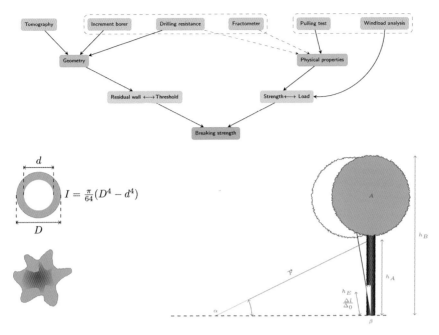

Figure 9.6 The assessment of the stem of a tree is either based on the distribution of sound wood, or on an estimate of the load-bearing capacity of the stem. Tomography shows properties of a cross-section in two dimensions, increment borer and penetrometers (drilling devices) in one. A fractometer can estimate wood properties of small samples extracted with an increment borer. The distribution of sound wood in the cross-section is usually evaluated on the basis of the second moment of inertia, I (bottom left, D: outer diameter, d: diameter of the cavity). In real-world cases, the dimensions of D and d are highly uncertain. Static pulling tests (bottom right) monitor the reaction of a tree (strain and compression of the outermost wood fibers and tilt of the root plate) while the tree is loaded with a winch. The estimated load-bearing capacity is compared to the estimated design wind load.

Figure 9.7 Sample of an *Aesculus hippocastanum* with wetwood, extracted with an increment borer.

Extracted wood cores (Figure 9.7) can provide samples for the identification of fungi and for dendrochronological studies. These can help, for example, to date the start of the decline of a tree, and thus to infer its causes. These cores have to be extracted precisely, horizontally and radially. Meticulous documentation of the exact placing of the drill or auger, including height at the stem, angle to the horizontal and compass direction, is important. Often, the look and smell of wood cores is the best way to convince lay people of the results of an expert assessment.

Their relative ease of application, even from an aerial lift or while climbing the tree, has made tools that record the resistance to drilling the most widespread among the high-tech instruments used in tree assessment. They typically use a needle that is 1.5 mm in diameter, with a tip of 3 mm, and are less invasive than an increment borer.

When the drill path is radial and perpendicular to the stem axis, results can be compared to species-specific density patterns to locate decay and cracks. The more the path deviates from the radial-perpendicular direction, the more complicated will be the interpretation of results. In soft wood and cavities, the needle can diverge significantly from a straight path. In some species with high intra-ring density variation, some drills can record ring width. However, these data can only be evaluated quantitatively if the drill is perpendicular to the stem axis and is aimed at the pith. Although resistance to penetration correlates with dry wood density, there is no straightforward way to relate the measurements to wood strength because, especially in the early stages of decay, strength is reduced much faster than density. Even without defects, there is only a weak correlation between green density and strength. While cavities are easily detected, locating decay and wet wood will require experience and training.

There are devices that directly measure the strength of wood extracted with an increment borer but, because of the way the samples are taken, their small size and the high variability of the physical wood properties, their results are unlikely to be representative of a significant part of the stem (Matheny *et al.*, 1999).

Depending on the type of decay, the tree species, the form of the stem and the decay and the device used, the uncertainty of results from drilling devices can be high, typically in the range of a few centimeters with modern devices.

Drilling devices will breach defensive zones of the tree, and can allow decay to spread. Plugging the holes will not help, and the use of wood preservatives on plugs will increase the damage even more. Therefore, the number of drillings should be as low as possible.

Time of flight of stress waves across a stem cross-section increases, when the wood contains decay, cavities, cracks or included bark, because the stress waves travel around the obstacle. A variety of instruments is used to measure time of flight and the distance between start and end points of the signal. Today, most devices use several sensors simultaneously (Rust, 2000). These sonic tomographs (SOT) measure the time of flight of signals excited by hammer blows, and recorded with 8–24 sensors fixed to the wood around the stem. The sensors have to be in contact with the outermost growth ring.

From these data, apparent stress wave velocity is calculated, assuming a straight path of the stress waves, because their true path is not known. It is essential to measure the distance between sensors exactly, because any error in distance will cause an equivalent error in velocity.

Sound velocity in wood varies, depending on the angle to the main stem axis, because wood is an anisotropic material. Along the grain, sound velocity is several times higher than across the grain, and radial velocity is higher than tangential. Therefore, all sensors must be installed in a plane perpendicular to the stem axis. Otherwise, it is not possible to separate the effects of wood anisotropy and defects.

Because of the large variation of stress wave velocity within and between individuals of different tree species, most devices use relative stress wave velocity instead of the absolute velocity. To do this, all velocities might be scaled by the highest velocity measured in a cross-section. The result of sonic tomography is a map of apparent relative stress wave velocity, which will reveal areas of sound wood, as long as they can be traversed by stress waves on a more or less straight path from one sensor to another.

Electrical resistivity tomograms (ERT) display the distribution of resistivity across the stem cross-section (Rust *et al.*, 2007; Bieker and Rust, 2010a, 2010b; Bieker *et al.*, 2010). ERT has been applied to find discolored wood, ring shake and decay in trees, and even to detect roots. While stress wave tomograms of defect-free stems are more or less homogeneous, ERT can show strong species-specific patterns, even in sound trees. These correlate with features such as heartwood and wood pH. Three common patterns have been identified so far. Species with dry heartwood, like pines (*Pinus sp.*) will have tomograms with a central area of high resistivity. Species without regular heartwood, like beech (*Fagus sp.*) often have a homogeneous distribution of resistivity. The most complex are the oaks (*Quercus sp.*) and similar species, which have a homogeneous moisture distribution in their cross-section, but a strong pattern of pH and ion content, resulting in three concentric areas, differing in resistivity.

ERT needs point-like contacts to the outermost layer of wood, and can be applied to measure stems of any size non-destructively. Usually, 24 measurement points are used. The spatial resolution depends on the distance between neighboring sensors and the tree diameter.

Because sonic tomography and electrical resistivity tomography measure very different physical properties of wood, their combination can greatly improve the assessment of a tree (Rust *et al.*, 2007; Figure 9.8). Incipient decay is characterized by high stress wave velocities and a change in electrical resistivity. Cavities, decay, and cracks are difficult to distinguish in sonic tomograms. Combined with ERT, this becomes much easier, because cavities usually have a high resistivity and decay shows a low resistivity, while the wood around cracks often is unaltered (Figure 9.9). Where SOT often shows a large area of low stress wave velocity around a crack, which might be interpreted as cavity, ERT can show low resistivity, indicating the presence of sound wood around a crack.

While the methods covered so far look at some measure of strength of the stem or branch only, static pulling tests (Ylinen, 1953; Sinn and Wessolly, 1989; Wessolly, 1991) compare the load-carrying capacity of the stem with the estimated wind load of the tree. The design wind load is estimated from the projected area of the tree and a range of site factors, including the local wind speed, either Beaufort 12 (>32.7 m/s), or that with an exceeding probability of 2% (i.e., once in 50 years). Adjustments are made for the likelihood of resonance in the wind response of the tree, as well as for damping effects that will reduce the actual load a tree experiences in natural winds. The load-carrying capacity is estimated based on the reaction of the tree to a static load applied by a winch.

The strain (compression and/or dilation) of the outermost fibers is measured at points of the stem where, based on a visual inspection, it is expected to be largest. In cross-sections of similar size and form, this strain will be increasing with defect severity. The load test is stopped at a strain of typically 0.05%, which is entirely reversible and non-destructive.

Based on Hook's law, the linear relationship between load and strain is extrapolated to tabulated values of strain at primary failure, in order to estimate the load necessary to cause primary failure. These strains are in the range of 0.2–0.5%. This critical load is compared with the design wind load of the tree. In order to account for the many uncertainties involved in the measurements and estimates, the assessment is based on a worst-case scenario. Furthermore, the threshold ratio is higher than unity. If the ratio of critical load and design wind load exceeds 1.5, the tree is considered to be safe.

Strength data of green wood are available for many European and North American tree species. However, they are the results of tests using small, clear wood samples, which

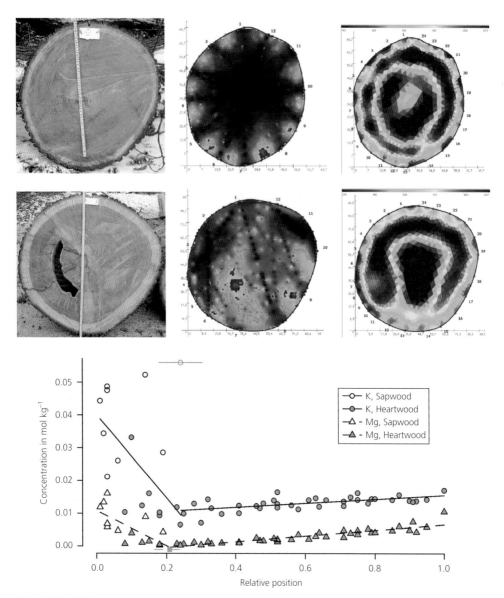

Figure 9.8 While SOT (top centre) depends on physical wood properties, ERT (top right) depicts wood chemical properties, such as ion concentration (bottom), as this example of two oak trees (top left) demonstrates.

overestimate the strength of stems, and are therefore adjusted with empirical factors. A recently developed method allows measuring the limit of proportionality (primary failure) in standing trees when pulled to failure (Detter *et al.*, 2013).

Pulling tests need special equipment and trained operators. To install the pulling rope in the tree, two climbers are needed for safety reasons. The evaluation of the data can be quite involved and, for these reasons, this method is usually among the most expensive.

Stress wave velocity	Electrical resistivity	Wood
high	high	intact
high	low	incipient decay wet wood
low	high	cavity/crack
low	low	decay

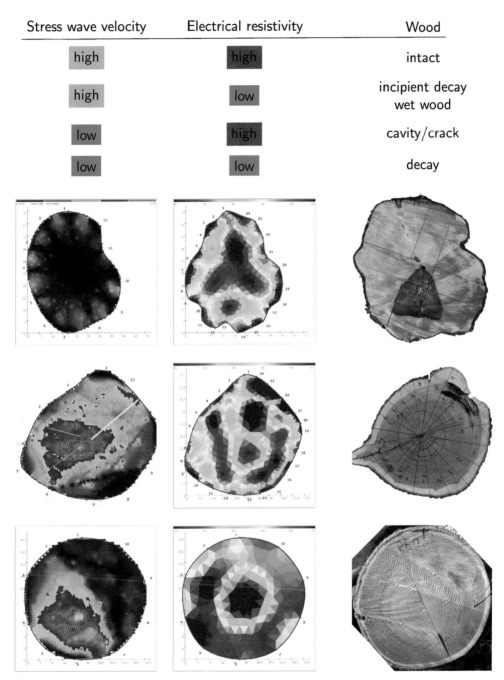

Figure 9.9 The combination of SOT and ERT can be used to differentiate between defects.

Another set of methods uses a combination of wind load estimates, tree dimension and measurements of internal decay (e.g., from tomograms) to assess the safety of trees. They are used in the form of a set of charts on paper, or as free or paid web services on the internet (Figure 9.10). The result is a factor of safety and, in the case of reduced

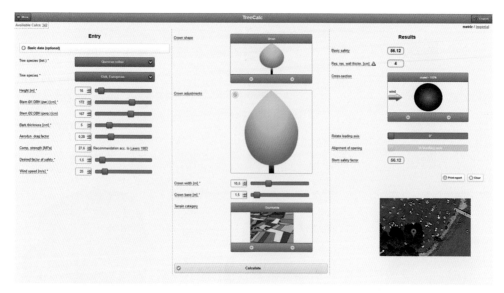

Figure 9.10 Screen shot of a web service for online structural tree analysis (Treecalc). (www.treecalc. com. Reproduced with permision of Arbosafe UG.)

safety, a recommendation of how much the tree should be reduced in height in order to achieve the desired factor of safety.

9.3.3.2 Root loss and anchorage

Static pulling tests can test the anchorage strength of trees. As described above, the design wind load is estimated and a load applied with a winch. Sensors monitor the increasing inclination of the root plate. Recent research has shown that there is a close correlation between the load required to incline a tree by 0.25° and the load required to pull it to failure. This load is estimated by extrapolation. Division by the design wind load gives a factor of safety, which is usually required to be above 1.5.

There is an ongoing debate whether static pulling tests are relevant to tree safety, because wind loads in nature are dynamic. A dynamically applied force can deflect or tilt a tree much more than a static force of the same size. Therefore, the pulling test method assesses the likelihood of root failure with regard to a static equivalent of the dynamic effects of natural winds on a tree.

The movement of the root plate of potentially unsafe trees can be monitored and analyzed to derive indicators of stability. In one system which is being offered commercially, trees are categorized according to the degree of basal tilt in high winds. Thresholds are 0.6° and 1.0° basal tilt. To date, scientific evidence on the validity of these thresholds is scarce, but our recent research indicates that this is the range where primary failure of the root system occurs.

Based on our recent research, dynamic measurements in high winds can be used to assess the risk of root failure. The strong correlation of root plate tilt and publicly available data on wind speeds can be used either to directly show that a tree did not tilt more than 0.6°, or to validate the estimated design wind load in static pulling tests (Figure 9.11).

Under favorable circumstances, roots starting from 2 cm in diameter and down to several meters in depth can be detected with ground-penetrating radar (GPR: Hruska *et al.*, 1999; Stokes *et al.*, 2002; Rust and Gustke, 2006; Bassuk *et al.*, 2011; Vianden, 2013).

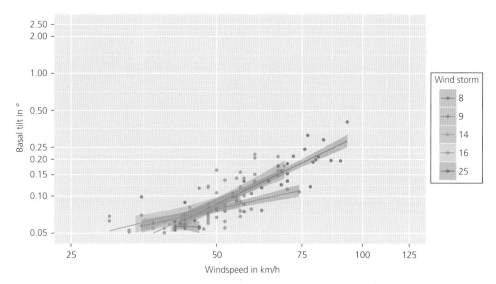

Figure 9.11 The correlation of root plate tilt and wind speed could be used to assess the safety of trees. Our recent research indicates that primary failure of roots starts above 0.5° (Rust and Detter, 2015).

Figure 9.12 Result of a GPR-scan of the root systems of three *Picea abies* (gray circles) in a park. Horizontal slice, showing structures that have been verified as roots by excavation (Vianden 2013. Reproduced with permission of Mitja Vianden).

To accurately locate roots, parallel scan lines should be no further apart than 15 cm. Two sets of perpendicular scan lines will increase accuracy further. In urban areas, three-dimensional analyses are very important because, in inhomogeneous soil, other reflectors (like rocks, cables, pipes, etc.) cause signals similar to roots, and there is a risk of misinterpretation. This can be minimized if detected objects have a spatial continuation and are connected to a tree. A thorough scan of a root system can take up to eight hours (Figure 9.12).

Root system imaging with GPR can be used to protect tree roots during excavation or construction, or to provide evidence that roots have been severed by such activities. Root collar excavation is an alternative. High-pressure air or water, or vacuuming, can be used to remove soil non-destructively.

Neither GPR nor root excavation can be used to assess the anchorage of a tree quantitatively, or to tell whether the roots are functional. Although sap flow measurements have been used to assess roots of urban trees (Cermak *et al.*, 2000; Rust and Gustke, 2001), this technology is too complex to be used on a routine basis. However, recent

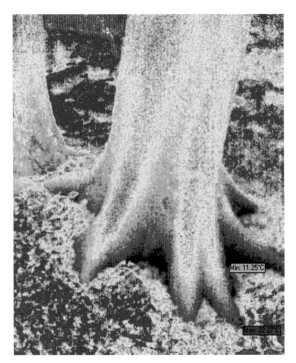

Figure 9.13 Thermographic images indicate sap flow at the base of a *Fagus sylvatica* (left) and an infection with *Pseudomonas syringae* in an *Aesculus* (right) (Rust, 2015).

studies have shown that thermographic images can reveal the flow of sap below the bark (Figure 9.13). A possible application is the assessment of damage caused by severing roots during construction.

9.3.3.3 Pathogens

Thermographic images can reveal the extent of some infections below the bark (see Chapter 6). Recent research showed that an infection of the stem of horse chestnut with *Pseudomonas* could be located non-destructively from a distance of several meters (Figure 9.13).

9.3.4 Risk categorization and reporting

Site, target, and tree result in the likelihood of failure and the likelihood of impacting a target. To categorize these likelihoods, they are entered into a likelihood matrix (Figure 9.14) to estimate the likelihood of a tree failure impacting a specific target within the assessment time frame. To rate the risk, a risk matrix can be used (Figure 9.14) that is based on the consequences of failure and the likelihood of the tree to impact a specific target. This is done for every likely failure mode of the tree. Because multiple risks in a tree cannot be aggregated mathematically, the tree might be designated the highest of its multiple risks.

The possible outcomes of this procedure are risk levels designated extreme, high, moderate, and low. The risk is extreme where failure is imminent, with a high likelihood of impacting a target, and the consequences of failure are severe. In such a situation, it should be recommended that mitigation measures should be taken as soon as possible. It may be necessary to restrict access to the target zone until the risk is mitigated.

Likelihood of failure	Likelihood of impacting the target: Very low	Low	Medium	High
Imminent	Unlikely	Somewhat likely	Likely	Very likely
Probable	Unlikely	Unlikely	Somewhat likely	Likely
Possible	Unlikely	Unlikely	Unlikely	Somewhat likely
Improbable	Unlikely	Unlikely	Unlikely	Unlikely

Likelihood of failure and impact	Consequences: Negligible	Minor	Significant	Severe
Very likely	Low	Moderate	High	Extreme
Likely	Low	Moderate	High	High
Somewhat likely	Low	Low	Moderate	Moderate
Unlikely	Low	Low	Low	Low

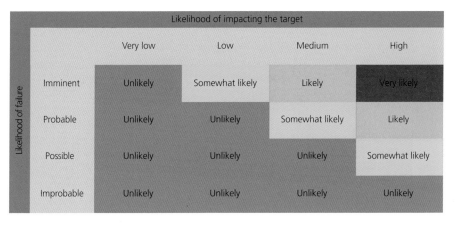

Figure 9.14 Matrix to assess the likelihood of a tree failure impacting a specific target within the assessment time frame (top) and matrix used to assess the level of risk (Dunster *et al.* 2013. Reproduced with permission of the International Society of Arboriculture).

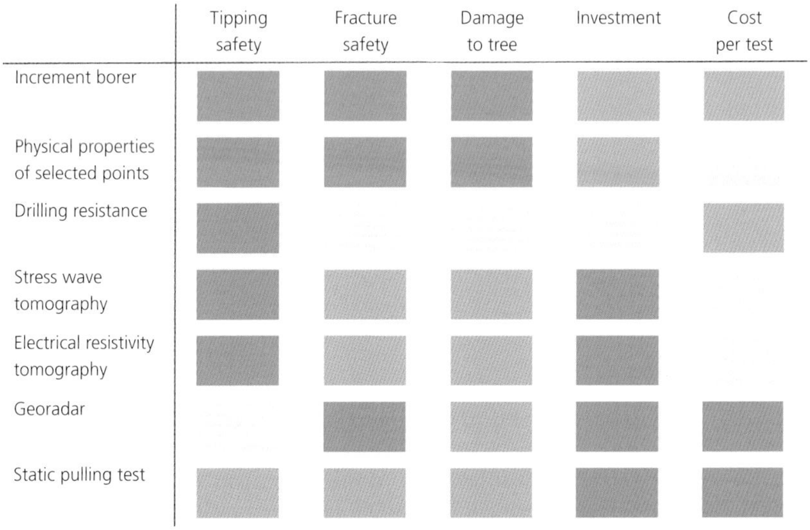

Tipping safety
Fracture safety
Damage to tree
Investment
Cost per test
Increment borer
Physical properties of selected points
Drilling resistance
Stress wave tomography
Electrical resistivity tomography
Georadar
Static pulling test

Figure 9.15 Comparison of methods in advanced tree assessment. Blue: good, yellow: fair, red: unfavorable.

When the consequences of failure are significant and impacting the target is very likely or likely, it is a high risk, as is the case if consequences are severe and occurrence is likely. The risk assessment report should contain recommendations for mitigation. The time frame for action depends on the risk tolerance of the owner.

For moderate-risk trees, the report might contain recommendations for mitigation, retention, or monitoring. As in high-risk situations, the time frame for action depends on the risk tolerance of the owner or manager.

The levels of risk resulting from an assessment of a population of trees form the basis for prioritizing mitigation or removal. Perception and acceptance of risk vary widely. The evaluation of risks identified by the tree risk assessor is usually done by the tree owner or risk manager, who decide on what, if any, actions should be taken, and on their timing.

Reporting will mostly take the form of written reports or records in a database. Written reports on a tree risk assessment can range from short memos to detailed reports. In any case, they should contain the site location, a means to identify the trees inspected, date of inspection, people involved, and results.

9.4 Conclusions

Legal requirements and economic pressure and knowledge of tree biology have raised the technical quality and forced the widespread use of tree inventories. In general, the risks from urban trees are low but, in many countries, health and safety legislation requires tree owners to have a suitable and sufficient risk assessment, and to apply measures that are reasonable and practicable.

Quantitative knowledge of the resource is the foundation of tree management. A tree inventory provides data such as number, location, size, and species to support negotiations on budgets for tree care, to allocate resources and to organize tree management. It provides a framework for the documentation of tree assessment, keeping records of tree inspections and maintenance.

Much progress has been made in recent years to formalize the way that trees are assessed. Two important institutions from the arboricultural industry – ISA in North America and FLL in Germany – have produced guidance on the different levels of tree assessment, risk categorization, and appropriate time frames.

Several technological innovations and their scientific evaluation help to assess trees in much more detail than just 20 years ago (Figure 9.15). Even so, quantitative estimates of the risk of failure are, with the possible exception of pulling test, not available today. More research is needed to fully utilize the level of detailed information made available by these devices.

However, an increased application of the principles outlined here, and detailed in several in-depth treatments elsewhere, could raise the standards of urban tree risk management substantially.

References

Bassuk, N., Grabosky, J., Mucciardi, A. and Raffel, G. (2011). Ground-Penetrating Radar Accurately Locates Tree Roots in Two Soil Media Under Pavement. *Arboriculture and Urban Forestry* **37**(4), 160–166.

Bieker, D. and Rust, S. (2010a). Electric resistivity tomography shows radial variation of electrolytes in *Quercus robur*. *Canadian Journal of Forest Research* **40**, 1189–1193.

Bieker, D. and Rust, S. (2010b). Non-destructive detection of sapwood and heartwood in Scots pine (*Pinus sylvestris* L.) *Silva Fennica* **44**(2), 267–273.

Bieker, D., Kehr, R., Weber, G. and Rust, S. (2010). Non-destructive monitoring of early stages of white rot by T*rametes versicolor* in *Fraxinus excelsior*. *Annals of Forest Science* **67**, 210.

Bond, J. (2006). Foundations of Tree Risk Analysis: Use of the t/R Ratio to Evaluate Trunk Failure Potential. *Arborist News* **23**, 31–34.

Cermak, J. Hruska, J., Martinkova, M., Prax, A., Cerm, J., Hru, J. and Martinkov, M. (2000). Urban Tree Root Systems and Their Survival near Houses Analyzed Using Ground Penetrating Radar and Sap Flow Techniques. *Plant and Soil* **219**, 103–116.

Detter, A., Rust, S., Rust, C. and Maybaum, G. (2013). *Determining strength limits for standing tree stems from bending tests*. In: Ross, R.J. and Wang, X. (eds). 18th International Nondestructive Testing and Evaluation of Wood Symposium, 226. USDA FPL, Madison, USA.

Dunster, J.A., Smiley, E.T., Matheny, N. and Lilly, S. (2013). *Tree Risk Assessment Manual*. International Society of Arboriculture, Champaign, Illinois.

FLL (2004). *Richtlinie zur Überprüfung der Verkehrssicherheit von Bäumen*. Bonn.

FLL (2010). *Richtlinien für Regelkontrollen zur Überprüfung der Verkehrssicherheit von Bäumen*. Bonn.

FLL (2013). *Baumuntersuchungsrichtlinien*. Bonn.

Hruska, J., Cermák, J. and Sustek, S. (1999). Mapping Tree Root Systems with Ground-Penetrating Radar. *Tree Physiology* **19**(2), 125–130.

Jacobs, M. (1936). The effect of wind on trees. *Australian Forestry* **1**(2), 25–32.

Kane, B., and Ryan, H.D.P. (2003). Examining Formulas That Assess Strength Loss due to Decay in Trees: Woundwood Toughness Improvement in Red Maple (*Acer Rubrum*). *Journal of Arboriculture* **29**, 209–217.

Kane, B.C.P. and Ryan, H.D.P. (2004). The Accuracy of Formulas Used to Assess Strength Loss due to Decay in Trees. *Journal of Arboriculture* **30**, 347–356.

Matheny, N.P., Clark, J.R., Attewell, D., Hillery, K., Graham, A.W. and Posner, G. (1999). Assessment of Fracture Moment and Fracture Angle in 25 Tree Species in the United States Using the Fractometer. *Journal of Arboriculture* **25**, 18–23.

Mattheck, C, Bethge, K. and West, P.W. (1994). Breakage of Hollow Tree Stems. *Trees – Structure and Function* **9**, 47–50.

McPherson, E.G. (2007). Benefit Based Tree Valuation. *Arboriculture and Urban Forestry* **33**(1), 1–11.

McPherson, G.E. (1992). Accounting for Benefits and Costs of Urban Greenspace. *Landscape and Urban Planning* **22**, 41–51.

Mergen, F. (1954). Mechanical Aspects of Wind-Breakage and Windfirmness. *Journal of Forestry* **52**(2), 119–125.

National Tree Safety Group. (2011). *Common Sense Risk Management of Trees*. NTSG Edinburgh.

Nielsen, A.B., Östberg, J. and Delshammar, T. (2014). Review of Urban Tree Inventory Methods Used to Collect Data at Single-Tree Level. *Arboriculture and Urban Forestry* **40**(2), 96–111.

Niklas, K.J. (1992). *Plant Biomechanics*. Chicago University Press, Chicago.

Östberg, J., Delshammar, T. and Nielsen, A.B. (2013). *Standards for Conducting Tree Inventories in Urban Environments*. Swedish University of Agricultural Sciences (SLU), Alnarp.

Rinn, F., Schweingruber, F.-H. and Schär, E. (1996). Resistograph and X-Ray Density Charts of Wood. Comparative Evaluation of Drill Resistance Profiles and X-Ray Density Charts of Different Wood Species. *Holzforschung* **50**(4), 303–311.

Rust, S. (2000). *A new tomographic device for the non-destructive testing of standing trees*. Proceedings of the 12th International Symposium on Nondestructive Testing of Wood. University of Western Hungary, Sopron, pp. 233–238.

Rust, S. (2015). Perspektiven für die Baumuntersuchung. In: Dujesiefken, D. (ed). *Jahrbuch der Baumpflege*, 168–179. Haymarket Media, Braunschweig.

Rust, S. and Detter, A. (2015). Weiterentwicklung der Zugversuchsmethode auf wissenschaftlicher Grundlage. In: *Contributions to Forest Science*, 78–97. Beiheft 17.

Rust, S. and Gustke, B. (2001). Ein neues Verfahren zum Nachweis von Wurzelschäden an Straßenbäumen. *Landschaftsarchitektur* **12**, 12–13.

Rust, S. and Gustke, B. (2006). Baumschutz am Deich. *Baumzeitung* **40**, 30–32.

Rust, S., Weihs, U., Günther, T., Rücker, C. and Göcke, L. (2007). *Combining Sonic and Electrical Impedance Tomography for the Nondestructive Testing of Trees*. International Symposium on Nondestructive Testing of Wood, University of Minnesota, Duluth, USA.

Sinn, G. and Wessolly, L. (1989). A contribution to the proper assessment of the strength and stability of trees. *Arboricultural Journal* **13**(1), 45–65.

Stokes, A., Fourcaud, T., Hruska, J., Cermak, J., Nadyezdhina, N., Nadyezhdin, V. and Praus, L. (2002). An evaluation of different methods to investigate root system architecture of urban trees *in situ*: I. Ground-penetrating radar. *Journal of Arboriculture* **28**, 2–10.

Tirén, L. (1928). Einige Untersuchungen über die Schaftform. *Meddelanden fran statens skogsförsöksanstalt* **24**(4), 7–23.

Vianden, M.J. (2013). *Non-destructive detection of tree roots with geophysical methods in urban areas* (in German). PhD Thesis, University of Göttingen, Germany.

Vogel, S. (1994). *Life in moving fluids*, 2nd edition. Princeton University Press, Princeton, New Jersey.

Wagener, W.W. (1963). *Judging Hazard from Native Trees in California Recreational Areas: A Guide for Professional Foresters*. US Forest Service Research PSW-RP-1.

Weltecke K. and Gaertig, T. (2012). Influence of soil aeration on rooting and growth of the Beuys-trees in Kassel, Germany. *Urban Forestry & Urban Greening* **11**(3), 329–338.

Wessolly, L. (1991). Verfahren zur Bestimmung der Stand- und Bruchsicherheit von Bäumen. *Holz als Roh- und Werkstoff* **49**, 99–104.

Ylinen, A. (1953). Über die mechanische Schaftformtheorie der Bäume. *Holz als Roh- und Werkstoff* **11**(6), 16–17.

CHAPTER 10

Tree preservation, maintenance and repair

Steffen Rust

HAWK Hochschule für angewandte Wissenschaft und Kunst, Fakultät Ressourcenmanagement, Göttingen, Germany

10.1 Introduction

Most sites where urban trees are planted and grow are very different from their natural rural habitat. They need technical solutions, maintenance and protection to establish and survive, often under harsh conditions. Mortality is high and life expectancy low among urban trees (Nowak *et al.*, 1990, 2004). In times where the benefits from trees in urban areas are increasingly recognized and expressed in monetary terms (Nowak *et al.*, 2008) but, at the same time, tree budgets are often reduced, efficient use of resources is paramount. A thorough understanding of the solutions available is therefore required.

10.2 Preserving existing trees during development

The ecological benefits of trees grow with their size and, in many cities, mature trees significantly increase the monetary value of a property. Even so, trees can be severely damaged or destroyed during construction. There are many activities that could impact on trees, including trenching, lifting or laying of hard surfaces, demolition, installing services, general building, driving of heavy machinery, and landscaping close to trees.

Because trees are often protected by legislation from local to international level, local planning authorities can issue tree preservation orders, and leverage regional, national, or international conservation legislation to protect trees, for example as habitats for endangered species.

Several countries have guidelines and standards that formalize the role of tree preservation in development. In Germany, the RAS-LP4 ("Guidelines for the construction of roads – Part landscape management, section protection of trees, vegetation, and animals during construction" – FGSV, 1999) gives a detailed pictorial and technical account of possible problems and solutions. In the UK, the British Standard "Trees in Relation to Design, Demolition and Construction to Construction – Recommendations" (BS 5837: 2012) details the organizational steps that should be taken to ensure that trees are appropriately and successfully retained when a development takes place. It describes tree constraints plans, tree protection plans and arboricultural method statements.

Trees are very vulnerable on development sites, and may be affected either immediately (if removal or pruning is necessary to accommodate a development), or in the longer term. This may result from disturbance during the development process, or even some

Urban Tree Management: For the Sustainable Development of Green Cities, First Edition. Edited by Andreas Roloff.
© 2016 John Wiley & Sons, Ltd. Published 2016 by John Wiley & Sons, Ltd.

time later, following pressure to remove or prune trees from the occupants of new buildings. The design layout should take these issues into account.

Once it has been decided which trees are to be incorporated into a design layout (see Chapter 2), it is important to ensure that they will survive the development process. Tree officers need to be involved as early as possible in the planning process. Expert arboricultural advice should be obtained in the early stages of a project, as well as during development and implementation.

10.2.1 Tree constraints plan

For the survival of trees during construction, the first and most vital stage, in the design and layout of any site, is the creation of a tree constraints plan, as described in the BS 5837 (or similar measures in other countries, according to national guidance. In the following section, the process is described by the example of the British Standard). The early availability of a tree constraints plan to the development design team will benefit the developer by:

• reducing redesign time, as trees will be identified before the initial design is drawn up;
• reducing the risk of applications not being registered by development control;
• reducing the risk of tree issues being raised during the planning application process;
• increasing the speed at which tree issues can be dealt with during the application process.

The tree constraints plan should be a combination of the information gathered during a topographical survey (location of all trees, shrubs and hedges and other relevant features such as streams, buildings and spot level heights) and an accurate tree survey.

10.2.2 Tree survey

Prior to development, all trees on the site should be mapped and categorized. Plans should include precise locations of the trunk, crown, and roots. Although mapping of roots is technically feasible, most planners will resort to allometric relationships (see below).

Not all trees will be of equal quality and suited for retention. Therefore, trees should be categorized, identifying those individuals that can, and should, be retained. Examples of unsuitable trees could be those that are in irreversible decline, have a serious structural defect, or trees that will become a hazard after losing shelter from other trees or buildings. These trees might, however, have conservation value, and might be preserved for those reasons.

According to BS 5837 (2012), the tree survey should contain the following information about each tree on the site that has a stem diameter above 75 mm, measured at 1.5 m above ground level, and those trees of smaller diameter that are of particular interest or potential value:

• A tree reference number (this should relate to the tree constraints plan).
• Tree species.
• Height.
• Stem diameter taken at 1.5 m from ground level (diameter at breast height).
• Branch spread (in four directions north, east, south and west).
• Height of crown clearance above ground.
• Age class.
• Physiological condition.
• Structural condition.

- Preliminary management requirements.
- Estimated safe useful life expectancy.
- Category grading.

It is important to remember that many people involved in development have wrong concepts about the location of roots. It may help to instruct them, perhaps referring to the drawings in RAS-LP4, that the parts of a tree that lie below the soil surface – its roots – are just as important as those above ground, and that every effort should be made to ensure that the roots of retained trees are not damaged during the construction process. The health of trees with root damage often deteriorates, resulting in the need to remove the tree, or even in its structural collapse.

Tree roots can be easily damaged by (Figure 10.1, and also see Chapter 4):

- abrasion;
- crushing by vehicles/plant equipment and/or storage of building materials or soil;
- compaction of the surrounding soil, leading to root death by asphyxiation (lack of oxygen) or drought;
- severing and removal of roots by excavation;
- poisoning from spillage or storage of fuel, oil, chemicals, etc.;

Figure 10.1 Often, protection of trees during construction is neglected. Trees suffer from compaction and direct damages to roots, stem, and branches.

- changes in soil levels around trees, resulting in root death as a result of exposure or asphyxiation;
- installation of impermeable surfaces, leading to a decline in tree health due to lack of water.

It is vital therefore that the tree constraints plan clearly shows the root protection area (RPA) of each tree.

10.2.3 Root protection area

The root protection area (RPA) is the minimum area around a tree containing the roots and rooting volume, where the soil structure and the tree need to be treated as a priority. Because roots are rarely mapped prior to development, estimates based on allometric relationships are used in most cases. Typically, the stem diameter at some standard height (1–1.5 m) is multiplied by a factor of about 12 to calculate the radius of a circle around the stem base as RPA (BSI, 2012; FGSV, 1999). German guidelines recommend either a radius of 2.5 m or the result of the above calculation, whichever is larger (FGSV, 1999). For a multi-stemmed tree, this radius is ten times its basal diameter measured above the root flare.

10.2.4 Tree Protection Plan

A Tree Protection Plan is an essential aspect of tree protection during development. It is a scale plan, showing:
- any proposed or existing buildings or structures;
- all retained trees, both on and neighboring the site, and their corresponding root protection areas and crown spreads (N, E, S and W);
- the location of protective fences or barriers (details of how these are to be constructed must be supplied);
- proposed location of all plant and materials storage;
- drainage runs, roads and driveways;
- existing and new accesses;
- any other surface or underground features that may affect the trees on, or neighboring, the site.

10.2.5 Arboricultural method statement

Where proposed developments or site activities might have an adverse impact on retained trees, an arboricultural method statement may be essential, in order to implement controls to safeguard those trees throughout the construction process.

A method statement is a document detailing how a particular process will be carried out. In the arboricultural industry, such a statement is commonly used to describe how construction works can be carried out close to trees. It should detail the methodology for the implementation of any aspect of development that has the potential to result in the loss of, or damage to, a tree, and explain how this damage will be avoided. Local planning authorities can request this document as a condition of planning consent.

The method statement contains a timetable indicating when, and how, specific works adjacent to trees should be carried out. This will cover items such as the installation of protective fencing, hand excavation within tree protection zones, or surface changes. Engineering specification sheets should be included for items such as the design of protective fencing, special surfaces, or methods of trenching. Bills of quantities for materials are to be included where necessary. Site supervision by an arboriculturist may be stipulated for some, or all, of the operations associated with trees. Experience shows that regular on-site supervision of tree protection measures during the construction phase is essential.

Items to include within a method statement:

• Schedule of tree works, such as pruning prior to and upon completion of construction works.
• Tree protection zone (TPZ) (distances, type of fencing).
• Specification for surface changes.
• Specification for level changes.
• Trenching methods.
• Location of bonfires, chemicals.
• Contingency plans (chemical spillage, collision, emergency access to the TPZ).
• Post-construction landscaping near trees.
• Tree planting (storage of trees, site preparation).
• Contact listing (council, arboriculturist, architect).

A solid fence, firmly fixed in the ground so that it cannot be easily moved or opened, is often the best way to protect trees. When access to the root protection zone cannot be avoided, compaction can be reduced by placing steel plates on a thick layer of mulch or gravel. Geotextiles allow easy removal.

10.2.6 Pre-development treatments

Often, trees designated for retention will need some form of treatment in order to reduce the impact of construction. Typically, this is access facilitation pruning or root ditches.

10.3 Maintenance of planted and established trees

Urban trees grow in hard landscapes (TDAG, 2014), sometimes very different from their natural habitat, and often require lifelong maintenance. After planting, without the shelter from neighboring members of their cohort in a gap of a naturally regenerating stand, they may need stakes or other means to stay upright. Often, they need protection from vandals or solar radiation, and a supply of water, fertilizer or mulch in order to grow. Formative pruning is an essential part of the post-planting care, as remedial pruning later in the tree's life will be much more expensive and might be harmful to the tree. Regular monitoring, adjustment and, eventually, removal of support and protective devices will prevent damage to the trees. Later in their lives, mature trees with decay caused by collisions of cars, pruning wounds, or construction nearby might need cables or pruning to maintain their structural integrity.

10.3.1 Physical support

Trees may need physical support at any stage of their life to reduce the risk of failure of branches, stems, or root systems.

10.3.1.1 Support systems for trees after planting

Excessive movement of the root ball hinders the growth of roots out of the root ball into the surrounding soil and, thus, the establishment of young trees. Many young trees are able to stand upright alone after planting. However, most are staked after planting, not only to keep them upright, but also to protect them against mowing equipment and vandals. In regions where wind from one direction predominates, support systems will

Figure 10.2 Young trees can be supported by various methods. In this case, stakes, tree shelters and grazing protection help to establish trees.

prevent trees from becoming slightly tilted during their first years after planting. In some situations, tree shelters will be advantageous, as can be an underground fixation of the root ball (Figure 10.2).

10.3.1.2 Staking

Traditionally, the stem of young trees is supported by stakes. Stakes should be driven into the ground outside of the root ball to avoid damages to the root. The connection to the tree should be broad, smooth, elastic, and non-abrasive. Hose-covered wire can easily damage the tree when the hose degrades. The tree should not be held rigidly but should, rather, be allowed some movement.

Recent studies have shown, that the mechanical support provided by high stakes can be very low (Appleton *et al.*, 2008; Eckstein and Gilman, 2008; Brehm, 2013). Furthermore, high staking results in smaller root systems and more slender stems – even reversed taper. Two low, 50–100 cm high stakes are mechanically and biologically advantageous. In windy areas and for larger trees, three or four stakes are required. Palms often need no extra support (see Chapter 12).

10.3.1.3 Guying

Trees can be supported by steel or fiber cables fixed to the ground, by another tree, or by some other external structure. Often, these cables run at an angle of 45° from the tree to a ground anchor. Care must be taken that the connection to the tree does not damage it. Hose buffers degrade with time, and then the wire can cut into the stem. Trees should be allowed to move, as this will increase the root system and they will grow less slender. Wires and cables should be visible to pedestrians and maintenance personnel, to avoid tripping and injuries.

10.3.1.4 Root ball anchorage

Stakes can be unsightly, especially when a great number of trees are planted at a site. Root ball anchorage is invisible, and it allows the tree to sway naturally in wind. Thus, the impact on its growth might be lower than that of stakes.

The root ball can be fixed below ground by various methods. These should ensure that roots are not damaged by wires, and that the fixation decays fast enough to avoid damage to the tree.

A different approach is a metal pin sometimes called an "artificial tap root". At one side, it is inserted into the root ball from below, while the other side protrudes into the planting pit and prevents the tree from turning. Two recent studies had mixed results on its effectiveness. In one study, roots were damaged by the device, while in another, anchorage strength was not increased compared to an unsupported control.

Tree shelters protect against animals and lawnmowers, and can act like miniature glasshouses. With most species, they increase survival and height growth, albeit at the cost of more slender stature. They are best suited where animals and competing vegetation are a problem – for example, in extensive plantings along highways.

Tree support has to be maintained regularly, and must be removed in time to prevent girdling or abrasion. Although a common recommendation is to remove stakes after one year, trees might tilt in windy regions if not supported into the third year.

10.3.1.5 Support systems for tree crowns

Tree crowns with poor structure, such as weakly attached branches or co-dominant stems, and defects like decay or cracks, can be supported mechanically by bracing or cabling. In some exceptional cases, this needs to be combined with pruning to reduce the load in the supported parts (see Chapter 11). These installations will not be possible in all trees. When installed, they need inspection, maintenance, and regular replacement and repositioning. While corrosion and UV radiation weaken the support systems, secured stems and branches grow in size and weight. The guidelines of the FLL in Germany require eight years of design life; some products are guaranteed to last 12 years or even more.

10.3.1.6 Cabling and bracing

Cabling systems are installed either to prevent the failure of parts of the crown, or to hold limbs within the crown if they fail.

For many years, steel cable wires were used; today, however, non-invasive textile systems provide several advantages over traditional cabling. These are anchored with a sling around the tree. There is no need to drill through the wood, which can be an advantage in the case of already decayed wood, where the spread of fungi would be accelerated, or in poorly compartmentalizing species (Figure 10.3). Dynamic textile systems allow the tree to move and to adapt to increasing load. They restrict the movement to a secure range, and stop it more smoothly than steel wire.

Cables have to be repositioned periodically, or an additional system is needed when the secured limbs grow in length and weight. Sling systems avoid the accumulation of wounds over the years.

Before the installation of cables, the owner of the tree must be notified that the tree will need periodic inspection of condition, position, cable tension, and the tree's structural integrity.

Figure 10.3 Cables installed as a dynamic, non-invasive sling system (left). Cables have a finite design life, and trees grow in size. Therefore, new cables have to be installed periodically. In this tree (right), there are three generations of cables, both intrusive and non-invasive.

Cables can secure limbs with decay or cracks, long and slender branches, or co-dominant stems against failure. For this purpose, they should be installed horizontally at roughly two-thirds of the leader's length above the point where failure could occur. The stem that is supposed to secure an other stem must be able to carry the additional load. This load can be reduced, when the cable is installed with a slightly downward tilted angle, that is, lower on the securing stem than on the secured stem.

In most cases, a dynamic installation is used. This allows the tree to sway in low wind and to adapt to increasing loads, but will stop excessive movements smoothly. However, there are contradictory reports on the differences in adaptive growth between trees with static, compared with dynamic, systems.

A dynamic cabling system will reduce the swaying of limbs. This could either increase the stability of the tree as a whole, because it avoids resonance, or it could decrease it, because less energy is dissipated within the crown, so that it is directed towards the root system. Recent studies have reported ambiguous results with respect to whether cabling systems are likely to increase the load on the stem and the anchoring root system.

Static systems are used when a stem has already partially failed, or the development of a crack between co-dominant stems is likely, or when an old steel cable is replaced. Steel cables are anchored with non-invasive slings, or with traditional eye bolts or threaded rods.

One direct cable reduces sway in one direction only. In most cases, it will be better to use a triangular system or a combination of triangular installations. If this is not possible, a hub and spoke system attaching radial cables (spokes) to a central ring (hub) can be used.

The purpose of another set of cabling systems is to hold a limb within the tree if it breaks. They must avoid the acceleration of the falling part, so they are installed as static systems and at an angle between vertical and 45°. Because the broken stem will fall as far as the cable is strained, the cables should withstand at least twice the weight of the secured part. Dynamic systems can be designed less strongly.

Figure 10.4 Two veteran trees, a *Betula* growing on a *Quercus*, are propped and cabled to reduce the risk of failure.

Bracing systems are rods inserted into the tree to reduce the risk of two or more leaders spreading further apart or moving sideways in relation to one another. They can be used, for example, when the trunk is split at the base of two co-dominant stems. They are installed at the branch union and, in large trees, multiple rods are placed below the attachment.

Bracing damages the tree, so it should be a last resort. In many cases, pruning and static cables might be sufficient to stabilize the tree, but any bracing of branch unions must be supported by cables higher up in the crown.

In some cases, neither guys nor cables can be used to stabilize a tree. In these instances, branches or trees can be supported by props. The base must be installed in the ground without damaging the root system, but it must be strong enough to withstand the static and dynamic load in both directions. If possible, the top end should be attached to the tree non-destructively. This can be achieved with textile sling systems that restrict the upward and downward movement of the tree (Figure 10.4).

10.3.2 Protection against collisions

Collisions of vehicles and lawn mowing equipment are a frequent cause of damages in stems, low branches, and superficial roots. Often, wood-decaying fungi will enter the wood because of torn or destroyed bark.

Bollards or similar devices will prevent collision by, and compaction from, vehicles. Mulch, or other hard to walk on material, will reduce compaction by pedestrians. Flowerbeds or mulch around the base of the tree will eliminate the need to approach the tree too closely for mowing or string trimming.

10.3.3 Solar radiation

Solar radiation can kill the bark of thin-barked trees (see Chapter 6). This happens in young trees after planting, or when older trees or parts of them are suddenly exposed by the removal of neighboring trees and buildings, or parts of their own crown. Later, wood-decaying fungi colonize these areas.

Plastic or textile covers (e.g., jute) can even increase the temperature of the bark. Rush mats can reduce the heat load of tree stems considerably. One disadvantage can be that they hinder the monitoring of the stem for insects.

Some years ago, a special flexible white paint was introduced which reduces heat load on the stem and gradually degrades within 5–10 years. During this time, the tree can adapt to the increasing radiation reaching the bark.

10.3.4 Wound treatment

Wounds are caused by collisions of vehicles, lawnmowers, lightning strikes or pruning, to list just a few. They expose the wood and, if they are larger than a few square centimeters, will inevitably be colonized by wood-decaying fungi (see Chapter 11). For decades, they have been treated in various ways to minimize damage. However, there is no cure once wood starts to decay.

Decades of research have shown that wound dressing does not stop decay. Many additives, including fungicides, alcohol or wood preservatives, even increase the damage, because they kill living tissue around the wound without preventing decay. Often, persistent wound dressing conceals intensive decay. Wound dressing can prevent the cambium from drying and, to this end, can be applied to the edge of the wound, or to the entire wound if it is small.

When the bark is torn from wood, it can be put back into place and fixed. When the bark has been destroyed, the growth of a wound callus, starting from living ray cells at the surface of the wood, can be encouraged. In both cases, the affected area should be wrapped with opaque PE film within one or two weeks.

Cleaning, sterilizing and filling cavities are all harmful to the tree. Any attempt to remove wood infected with decay fungi will most likely breach the defensive zones of the tree and allow fungi to colonize formerly healthy wood. Likewise, the installation of drain tubes exposes healthy wood and may speed up decay (Figure 10.5). Shaping wounds does not help to close them. Any living bark should be left in place, while dry and dead bark can be removed.

Trees can be severely damaged by lightning. Often, long strips of bark and cambium are destroyed. Sometimes, the entire stem is split in half. The extent of the damage cannot be assessed immediately, especially when the root system is affected. Immediate actions should be limited to ensuring the safety of the tree, and minimizing further damage. Loose bark can be tacked onto the wood and covered with PE film.

10.3.5 Water management

In many parts of the world, demand for water is higher than supply, at least for parts of the growing season of urban trees. This is exacerbated by low infiltration rates due to paved or compacted planting sites, low soil volume, poor soil structure and the urban heat island. Lack of water is one of the major causes of failure of newly planted trees (see Chapter 5). Where this is likely, a detailed water management plan should accompany planning for planting.

Excess water can be equally damaging to trees, so sufficient drainage should be provided. Where storm water is a problem, planting sites can be designed to collect, store, and drain it. Storm water, collected from buildings and paved areas, might be available to irrigate trees.

Figure 10.5 *Laetiporus sulphuraeus* growing in a cleaned and shaped wound that has been treated with wood preservatives and wound dressing.

10.3.5.1 Planting site size

Irrigation may not be a reliable source of water. Especially during severe droughts, it may be restricted in order to conserve water for higher priority uses.

Wherever possible, tree roots should have access to sufficient soil volume that stores and provides water during rainless periods. The larger the tree grows, the less likely it will be that the original planting pit is of sufficient size. Therefore, trees need access to rootable volume in addition to their pit. Rough guidelines exist, recommending 0.75 m³ of soil for each square meter of crown projection area, or two cubic feet of soil for each square foot of canopy. A tree with a crown of 10 m diameter will require access to a soil volume of 9 m × 9 m and 1 m deep. Many planting sites are much smaller (Figure 10.6). A high soil water storage capacity is especially important in winter rainfall climates, or in tropical climates with a short, intense rainy season.

10.3.5.2 Irrigation

In most parts of the world, irrigation is essential, at least in the growing seasons after planting a tree. Several models are available to estimate the required amount of water. These range from fixed amounts of 30–100 liters per week and tree, as is regularly recommended in Germany, to sophisticated bioclimatological models.

Such models require a reference evapotranspiration, often locally available from agricultural services, and modifiers that take into account landscape, vegetation structure and plant species. To avoid over- or under-watering, irrigation systems can integrate

Figure 10.6 A small and compacted planting pit can cause even expensive planting stock to die (left: before planting, right: dead tree in this pit three years later).

information from sensors measuring soil water content or plant drought stress. However, data on soil moisture content from sensors installed in the vicinity of the tree might not reflect the true water status of the tree, because the root system is distributed across a large and heterogeneous volume.

To avoid competition with superficially rooting herbaceous vegetation, such as turf grass, the amount of water given at any one time should be sufficient to moisten the entire rooting space.

Irrigation basins, semi-circular bunds or micro catchments might help in irrigation and rain harvesting, by directing water to and holding it at the base of the tree.

Proactive irrigation in periods of high water availability may store water in soil layers accessible to trees, that they can utilize in periods of drought. Care must be taken to use water suitable for irrigation, to avoid salinity and related problems.

Bags or plastic rings that slowly release water at the base of the tree can reduce the frequency of irrigation. As with all plastic material in the landscape, they should not be allowed to disintegrate in place.

In most cases, trees are watered manually. It has proven useful to ensure that watering by contractors is delivered as planned. Projects involving the commitment of local neighborhoods to water newly planted trees can successfully reduce the cost of maintenance, although they, too, need monitoring. Unsupervised manual or automatic watering, especially when its main intention is to water turf grass, can cause hypoxia and superficial root growth in trees.

When irrigation systems are installed after planting, care must be taken not to damage the roots. Any trenching within the drip line will most likely sever roots.

10.3.6 Mulching
In their natural environment, often a forest, trees receive an annual layer of leaf litter but, in most urban areas, leaf litter is removed during maintenance. A well-established practice in many parts of the world is the application of a layer of material onto the soil surrounding the base of the tree. This is called mulching.

There is a wide range of intentions behind mulching. Mulch protects the soil from compaction, because often people avoid walking across it. Compacted soil might be loosened due to increased biological activity, most notably that of earthworms, underneath the mulch.

Mulch-suppressing vegetation has the effect of reducing competition for water and nutrients, giving a better appearance to the planting site, and eliminating the need to approach the stem base with mowing equipment, thus reducing the risk of widespread mowing damage.

Mulch might conserve soil water by reducing soil evaporation, supplying slowly released nutrients, and preventing soil erosion. It can also buffer extremes in soil temperature and reduce reflection of heat to the tree.

Although it is common belief, that, in general, mulching improves the site and the growth of trees, its effects may not always be beneficial. Examples for negative side effects are that rodents might be attracted to the root collar and damage it, especially when mulch covers the stem base, or that nitrogen can temporarily be bound in mulch with a high C/N ratio > 20–25.

Most often, coarse organic material is used, such as bark, wood chips, or leaves, but inorganic mulch like gravel, brick chips, or crushed coral might also be applied. It should be applied 10–15 cm thick, but a too thick, dense layer of mulch might impede gas exchange and water infiltration. Mulch should not be in contact with the stem. Organic mulches decompose and have to be replenished regularly (Figure 10.7).

Figure 10.7 Mulch should not cover the stem base as in this example. It must be replenished regularly, because it decomposes. Otherwise, roots that have grown into the mulch will be exposed and die.

Mulch from diseased plants should be composted at temperatures high enough to prevent pathogens from spreading to a new site.

10.3.7 Mycorrhizae

Mycorrhizae are mutualistic interactions between modified roots of a plant and the hyphae of a fungus, with the main purpose of nutrient transfer between the organisms (see Chapter 3, section 3.1.2). The fungal partner receives organic molecules from the plant, in exchange for providing hyphae as an enlarged absorptive surface for the plant. Mycorrhizal fungi contribute to soil development by improving the soil structure and mobilizing insoluble nutrients. Mycorrhizae can protect the plant, either directly or indirectly, against different forms of stresses. Usually, there is a high mortality when seedlings are planted on harsh sites and are not infected with mycorrhiza. Experiments with a wide range of plant species have shown, that under water deficit or salt stress, plants with arbuscular mycorrhizal fungi grow better than those without.

Compared with trees growing on natural sites, those on urban sites can have a lower species diversity and different species composition. There is a range of commercial products that can be used to add mycorrhizal fungi to urban soils. There are, however, almost no studies on the benefits of this practice, especially in mature urban trees, and results are inconsistent.

Before inoculating artificially, it should be checked whether the target site itself has the potential to infect the trees. In this likely case, it would be unnecessary to add inoculum. If there is a lack of mycorrhizal fungi, the causes should be elucidated and remedied. Artificial inoculation should use native fungi, even for introduced tree species, since these may adapt to native mycorrhizal fungi and form mycorrhiza with both native and co-introduced species. Very little is known about the invasion biology of mycorrhizal fungi.

10.3.8 Soil compaction

Soil compaction occurs intentionally or unintentionally for several reasons (see Chapter 4, section 4.1.2). A planting site might be compacted to achieve sufficient load-bearing capacity, or compaction may be caused by heavy machinery before planting during landscape construction, or by vehicles or pedestrians after planting the tree (Figure 10.8).

Compaction can increase superficial run-off, reducing the supply of water to the tree, can reduce gas exchange between soil and atmosphere, or cause water-logging in deeper horizons. In many tree species, roots will die in anaerobic conditions. Thus, root growth into compacted soil is reduced or stopped altogether, resulting in reduced vitality and safety of the tree (see Chapter 4).

Before trees are planted, rototilling or subsoiling can alleviate compacted soils. Around established trees, soil can be partially (e.g., in sectors) or totally replaced. Compacted soil is removed hydraulically or pneumatically, to avoid damaging the roots (Figure 10.9). Radial breakout trenches, reaching out from the edge of the existing root system and filled with amended soil, have been successful to some degree (Towbridge and Bassuk, 2004).

Over a period of several years, mulch may reduce compaction. However, studies on treatments involving the injection of compressed gas, water and, sometimes, soil amendments, or the installation of perforated pipes, found little or no sustained improvement.

Often, aeration systems like pipes or channels are installed at planting. Their design is mostly standardized, and does not account for soil and pipe properties. The pipes can become obstructed or overgrown quickly, and there are very few scientific studies on the efficacy of such systems.

Figure 10.8 Soil compaction reduces the permeability for water, carbon dioxide and oxygen.

Figure 10.9 Compacted soil can be removed pneumatically. Excavated roots are covered immediately with moist cloth.

10.3.9 De-icing salt

The use of de-icing salt (often sodium chloride, NaCl) on road surfaces and pavements nearby trees causes damage to trees by various mechanisms (see Chapter 4, section 4.1.2, and Chapter 6, section 6.2). In the soil, sodium can adversely affect structure by causing deflocculation, leading to soil compaction and reduced aeration. Sodium replaces other cations on their exchange sites, reducing nutrient availability. Sodium and chloride also lower the soil water potential, increasing the risk of water deficit to trees long after the application of salt. Leaves and bark are exposed to aerial spray from molten snow and ice blown into the air by traffic. Ions are taken up with sap flow and accumulate in leaves. All these effects together cause severe damage to, and frequently death of, roadside trees.

Although protective barriers can have a small local effect, the most effective way to reduce salt damage is to adjust quantities and techniques of salt application (e.g., using wet salt, or a mixture of sodium chloride and calcium chloride). In contaminated sites, the soil can be replaced, if necessary, with established trees in place. Salt-tolerant species should be selected where salt is a known problem.

10.3.10 Pruning to mitigate risk

Tree risk assessors frequently recommend pruning to mitigate risks (see Chapter 11). This includes crown raising to clear passages for pedestrians and vehicles, the reduction of limbs or the entire crown to reduce load, and the reduction of leaf area, where subsidence might be a problem.

10.3.10.1 Pruning to reduce load

Wind load increases in proportion to the area it acts upon, and even more so with wind speed. Because of crown and leaf reconfiguration, the increase with wind speed is not quadratic, as the theoretical formula would imply. However, reductions of height still reduce the wind load of a tree much more than raising the crown, where a similar amount of leaf area is removed, because of higher wind speed at greater height, and because any force at the top of the crown acts on a longer lever arm.

The reduction in height that is necessary to reduce the risk of failure of a tree to acceptable levels is, therefore, often low. Especially in light-demanding species, much of the foliage is concentrated in the periphery of the crown. Even with small reductions in height, the loss of leaf area might be substantial and much higher than relative reduction in tree height.

Pruning can change carbon allocation pattern and divert reserves from roots to regrowth of foliage. The consequences are reduced root growth, or even increased root mortality. There is a trade-off between leaf area and load reduction. In order to optimize the pruning dose, software tools are available that help to calculate the reduction of the crown necessary for a specified reduction in wind load (see Chapter 9).

Crown thinning will probably not reduce wind load as much as might be expected, because, while pressure drag is reduced, skin friction in the crown might actually increase. Often, the tree will quickly re-grow to the original leaf area.

10.3.10.2 Removing epiphytes

Epiphytes – plants growing on other plants – can range from symbiotic to parasitic (see Chapter 6, section 6.7). Examples of parasitic plants that are frequently managed on urban trees are mistletoes (*Santalales*) or dodder (*Cuscuta spp.*). Ivy is another example of an epiphyt that is frequently debated, because it is suspected that it can strangle or otherwise damage trees, increase their risk of wind throw, or obstruct risk assessment.

Mistletoes weaken their host and cause witch's broom and swellings of stem and branches. As long as there are only a few individuals in a tree, they can successfully be removed by pruning affected branches 10–40 cm below their insertion. Often, this treatment has to be repeated after 2–3 years to remove mistletoes that have been missed. In tree species with a low susceptibility and only a few mistletoes (5–10), one or two treatments can be sufficient. Removing mistletoes in highly susceptible tree species can be ineffective.

English Ivy (*Hedera helix*) is a liana that is native to Europe and North Africa, but invasive in North America. Studies, mostly in tropical forests, have shown that tree growth can be reduced when the canopy of a tree is covered by liana, and branch breakage and tree mortality can increase.

Evidence that English ivy increases the risk of wind throw is anecdotal. On the other hand, ivy provides habitat and food to many species, and at times when other sources are scarce. Therefore, in its natural range, there are good reasons to not remove ivy indiscriminately. When it is likely that ivy conceals a hazard during tree risk assessment, the tree inspector should request the removal of the liana, if and when this is allowed, according to laws protecting habitats and species.

10.3.10.3 Pruning to reduce transpiration

Building subsidence on shrinkable clay soils can be related to the transpiration of neighboring trees. Cyclic pruning to reduce tree water uptake has been recommended as a risk limitation strategy. Surprisingly, however, little is known about the effectiveness of this treatment. Crown thinning will be less effective than crown reduction, possibly because thinning exposes formerly shaded leaves to radiation. The reduction of leaf area has probably to be quite severe to have an effect; in one study, London plane trees were reduced 30% in height, equivalent to a 70% reduction of crown volume, and possibly even more in leaf area. This increased soil water availability for a maximum of three years (Hipps *et al.*, 2014), perhaps because of reduced transpiration or interception.

10.3.11 Ancient and veteran trees

Increasingly, the cultural, conservation, and scientific value of ancient and veteran trees is being recognized. An ancient tree is one that has passed beyond maturity and is old for its species, while a veteran tree is one with saproxylic habitat features which, while not yet necessarily ancient in years, possesses certain habitat qualities that are characteristic of an ancient tree (Fay, 2002).

These trees are often threatened by collapse, due to hollowing, or because pollarding stopped centuries ago and their trunks bear the weight of large pollard stems (Figure 10.10). Their treatment and protection need special consideration of the fact that sudden changes can be detrimental.

Based mainly on experience gained in the past two decades in the UK, there is ample guidance available on the treatment of ancient trees (Read, 2000; Lonsdale, 2013). Management plans for individual trees describe interventions for periods of, for example, 20 years.

10.3.12 Precautionary measures

Some precautionary measures are common to many areas of arboriculture. Utilities at and near a work site should be located, and potential conflicts taken into account. Regulations, for example, in Germany, may stipulate that contractors who plan to work

Figure 10.10 A legendary ancient oak as a destination in a national park. The soil is compacted from many visitors.

below ground must inquire for utilities on the site. Utilities need periodic maintenance and replacement, causing damage to trees above and below ground. Plans for tree planting should be adjusted if possible.

Before starting to work, a risk assessment of the workplace and the tree should be undertaken. Prior to climbing a tree, its suitability for climbing must be assessed.

A whole range of conservation legislation is related to trees. Plant and animal species living on and in trees, and trees as their habitat, may be protected by national and international policies. When such species or habitats are present, or are likely to be present, it might become necessary to consult experts and adjust the work planned.

10.4 Conclusions

Improved understanding of tree biology and technical advances in arboriculture have changed maintenance and repair of trees. Efficient protection during construction has been detailed in guidance available in several countries. Strict application of legal and technical options can greatly improve the survival of trees.

In harsh urban environments, adequate planting sites and post-planting care can significantly increase tree life expectancy. Solutions for most situations are available.

References

Appleton, B.L., Cannella, C.M., Wiseman, P.E., Alvey, A.A. (2008). Tree Stabilization: Current Products and Practices. *Arboriculture and Urban Forestry* **34**(1), 54–58.

Brehm, J. (2013). Untersuchungen von Baumverankerungen an Jungbäumen. In: Dujesiefken, D. (ed). *Jahrbuch der Baumpflege*, pp. 101–120. Braunschweig: Haymarket Media.

BSI Group (2012). *BS 5837: Trees in relation to design, demolition and construction. Recommendations.* London.

Eckstein, R., Gilman, E.F. (2008). Evaluation of Landscape Tree Stabilization Systems. *Arboriculture and Urban Forestry* **34**(4), 216–221.

Fay, N. (2002). Environmental arboriculture, tree ecology and veteran tree management. *Arboricultural Journal* **26**(3), 213–238.

Forschungsgesellschaft für Straßen- und Verkehrswesen (FGSV) (1999). Richtlinien für die Anlage von Straßen – Teil: Landschaftspflege: Schutz von Bäumen, Vegetationsbeständen und Tieren bei Baumaßnahmen (RAS-LP 4). Köln.

Hipps, N.A., Davies, M.J., Dunn, J.M., Griffiths, H. and Atkinson, C.J. (2014). Effects of two contrasting canopy manipulations on growth and water use of London plane (*Platanus x acerifolia*) trees. *Plant and Soil* **382**(1–2), 61–74.

Lonsdale, D. (ed, 2013). *Ancient and other veteran trees: further guidance on management.* The Tree Council, London, 212 pp.

Nowak, D.J., McBride, J.R., Beatty, R. (1990). Newly Planted Street Tree Growth and Mortality. *Journal of Arboriculture* **16**, 124–129.

Nowak, D.J., Kuroda, M., Crane, D. (2004). Tree Mortality Rates and Tree Population Projections in Baltimore, Maryland, USA. *Urban Forestry & Urban Greening* **2**, 139–147.

Nowak, D.J., Crane, D.E., Stevens, J.C., Hoehn, R.E., Walton, J.T., Bond, J. (2008). A Ground-Based Method of Assessing Urban Forest Structure and Ecosystem Services. *Aboriculture & Urban Forestry* **34**(6), 347–358.

Read, H. (2000). *Veteran trees: a guide to good management.* English Nature, Petersborough. 176 pp.

Towbridge, P.J., Bassuk, N.L. (2004). *Trees in the urban landscapes.* Wiley, New Jersey.

Trees & Design Action Group. (2014). *Trees in Hard Landscapes – A Guide for Delivery.* London.

Weltecke, K., Gaertig, T. (2012). Influence of Soil Aeration on Rooting and Growth of the Beuys-Trees in Kassel, Germany. *Urban Forestry & Urban Greening* **11**, 329–338.

CHAPTER 11

Tree pruning: Methods and parameters

Ulrich Pietzarka

Technische Universität Dresden, Tharandt, Germany

11.1 Introduction

Basically, every pruning of living parts of a tree has effects on the whole system "tree", which stimulates a number of different reactions by the plant. It is necessary to estimate these reactions in order to evaluate the complete pruning procedure, regarding its specific goals and objectives.

Therefore, an overview on the various reaction mechanisms of trees to pruning and their time scale is presented here. The effects of different parameters of pruning are discussed on the basis of these mechanisms. They are fundamental for judging the success of a pruning procedure. This judgment is only possible regarding the specific objectives (Gilman, 2012), so weighting of single parameters might be different where pruning is part of tree hazard management and improving the structure, or part of fruit production in orcharding, or just improving the appearance of trees (aesthetics).

11.2 Consequences of pruning

Pruning causes wounds (Figure 11.1) and, with this, immediate infection – first with mould, and later by wood-decaying fungi. In addition to the objectives of pruning, fast and effective closure of wounds is regarded as crucial for success in pruning (Figures 11.2, 11.4). Thus, various processes of compartmentalization of wounds (Figure 11.3) need to be considered (see Chapter 3, section 3.2). A quite detailed synopsis is given by Dujesiefken and Liese (2008), presenting the CODIT-principle.

It is of particular interest for pruning that compartmentalization relies especially on energy-demanding active processes. These induce changes in physiology as well as anatomy (Barry *et al.*, 2000), which is why the duration and effectiveness of compartmentalization depends on the vitality of the tree and the time of pruning. Furthermore, there are species-specific differences in the amount of living tissue in wood (parenchyma) where compartmentalization is induced (Dujesiefken and Liese, 2008).

Finally, the wound is closed by wound wood (phase 4 regarding the CODIT-principle – Dujesiefken and Liese, 2008). This especially results from cleavage of cambium cells at the edge of the wound. Here, the cambium is laid open circular to ellipsoidal. Also, other living tissues may be involved, that regain their ability to cleave and form an undifferentiated callus. This so-called barrier zone is distinctly different from normal wood, both in

Urban Tree Management: For the Sustainable Development of Green Cities, First Edition. Edited by Andreas Roloff.
© 2016 John Wiley & Sons, Ltd. Published 2016 by John Wiley & Sons, Ltd.

Figure 11.1 Fresh wound.

Figure 11.2 Ample wound wood.

chemistry and anatomy, and more energy is required to produce it. It shows a higher content of parenchyma, but fewer vessels (Schwarze *et al.*, 1999). This fact underlines the special importance of parenchyma in compartmentalization and wound closure. If the wound is completely closed, and pathogens encapsulated, they lack oxygen for survival.

An even and fast wound closure by cambial growth requires activity of the cambium, which should also be well supplied by assimilates. This is only warranted within the vegetation period, and also relies on the vitality of the tree. Supply of the cambium also

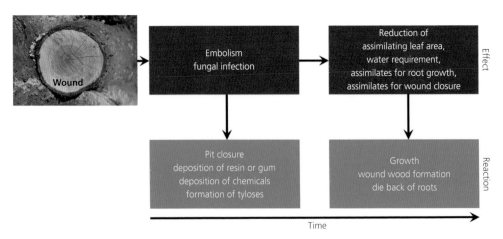

Figure 11.3 Effects of pruning and reactions of trees.

Figure 11.4 Decayed but effectively compartmentalized branch (left) and narrowly compartmentalized pruning wound (right) on hornbeam (*Carpinus betulus*).

depends on sap flow reaching the wound area that is substantially influenced by the pruning system. It is necessary to prevent so-called "supply shadows", where the flow of assimilates in phloem is hindered.

Apart from the time of pruning and the pruning system, compartmentalization also relies on the intensity of pruning and the amount of crown reduction. This not only determines wound size, and hence the energy and time necessary for its closure, but it influences the whole system "tree". Planning a pruning procedure it is therefore necessary to consider not only compartmentalization, but also later reactions of the tree.

Intensive pruning leads to an intense reduction of assimilating leaf area, at the latest in the next vegetation period. The water requirement is also reduced and the usually well-balanced root/shoot-relation is heavily disturbed. A critical shortage of roots with assimilates is possible, followed by reduced regeneration and growth of fine roots and, finally, even die-back of bigger roots. This is disadvantageous for structural integrity (certainly with a much lower crown mass), and opens up new entrances for pathogens which are difficult to detect in the root zone.

In summary, the impact of pruning on the root system has been poorly understood until now. The questions for a barely tolerable loss of crown area that does not lead to die-back of bigger roots can not be sufficiently answered. Also, the duration of this impact is relatively unknown. How many years does it take a tree to recover from a pruning procedure? We know from transplanting trees (Chapter 12) that this is age- and size-specific (Watson, 2005).

Finally, the development of new shoots following a pruning procedure needs to be taken into consideration. The tree aspires to adjust the disturbed root-shoot relationship by regenerating the lost part of the crown, or reduction of the root system. According to the pipe theory, there are also consequences for the conductive wood area (Roloff, 2004). The time and intensity of the development of new shoots depends on several factors. Besides species-specific factors, such as the ability to reiterate or growth rate, the vitality, the intensity of pruning and the time of pruning are also of special importance.

11.3 Important parameters

The consequences of pruning, and tree-specific reactions to it, have some important parameters that influence the time and effectiveness of compartmentalization and, therefore, the success of pruning procedures. These parameters may be divided into two groups (Table 11.1). While Pruning, the first group is no longer influenceable, but the second group combines a number of parameters directly linked to the pruning procedure.

It is necessary to evaluate these parameters while planning or executing tree pruning, and they are described below. The heavily discussed question of wound dressing might

Table 11.1 Important parameters influencing tree reaction after pruning.

Not influenceable	Influenceable
Ability to compartmentalize	Pruning system
Ability to reiterate	Intensity of pruning
Vitality	Date of pruning

Table 11.2 Ability of tree genera to compartmentalize wounds (Adapted from Dujesiefken & Stobbe 2002 and Gilman 2012).

Weak compartmentalizers		Effective compartmentalizers	
Aesculus	Malus	Acacia	Juglans
Acer (platanoides, negundo, saccharinum)	Myrica	Acer (rubrum, saccharum, campestre)	Lagerstroemia
Bauhinia	Paulownia	Albizia	Lysiloma
Betula	Peltophorum	Arbutus	Pinus
Brachychiton	Persea	Bucida	Quercus (sect. Quercus)
Celtis	Populus	Bursera	Robinia
Poinciana	Prunus	Castanea	Taxus
Eucalyptus	Quercus (sect. Lobatae)	Catalpa	Tilia
Erythrina	Salix	Carpinus	Swietenia
Ficus	Schinus	Cassia	Tabebuia
Fraxinus	Thuja	Fagus	Ulmus
Magnolia	Tsuga	Gleditsia	

be another factor, but is not part of pruning. As a general rule, most pruning cuts do not benefit from an application of pruning paint (see also Dujesiefken and Liese, 2008; Gilman, 2012).

The response capacity of trees for pruning is genetically determined, so the reaction will be individual, but also clone-, cultivar- or species-specific. This becomes apparent in the classification of tree genera in their *ability to compartmentalize* wounds (Table 11.2; Dujesiefken and Stobbe, 2002; Dujesiefken and Liese, 2008; Gilman, 2012; Lonsdale, 2013). This classification is not yet published for all species or genera relevant in urban areas. Also, a classification based on genus-level is not sufficient, because in some genera there are weak, as well as efficient, compartmentalizers (e.g., *Acer* or *Quercus*). In some cases, the classification is still controversial.

As previously described, the content of living tissue in wood (parenchyma), where compartmentalization is induced, is crucial for this estimation (Dujesiefken and Liese, 2008).

Figure 11.5 shows the content of parenchyma of several tree species (Frey-Wyssling and Aeberli, 1942; Wagenführ, 2007). For this purpose, the content of axial parenchyma and ray-parenchyma was summed, and weak and effective compartmentalizers marked, after Dujesiefken and Liese (2008) and Gilman (2012).

The average content is lowest for *Betula pendula* with 10.5%, and highest for *Morus alba* with 44.7%. It is obvious that weak compartmentalizers (marked red) show a significantly lower content of parenchyma than do more effective ones. Data prove that species of one genus may not always be classified the same way. For example, white oaks (*Quercus* sect. *Quercus*; here *Quercus robur* and *Q. alba*) have a higher content of parenchyma than red oaks (*Quercus* sect. *Lobatae*; here *Quercus rubra*). This corresponds to the classification of these species (Dujesiefken and Liese, 2008).

Figure 11.5 also shows that a classification considering only parenchyma content does not always reflect the experiences of practitioners. For example, the genus *Juglans* is classified as effective compartmentalizers (Gilman, 2012), but *Juglans nigra* has relatively low parenchyma content. A contrary example is the common ash (*Fraxinus excelsior*).

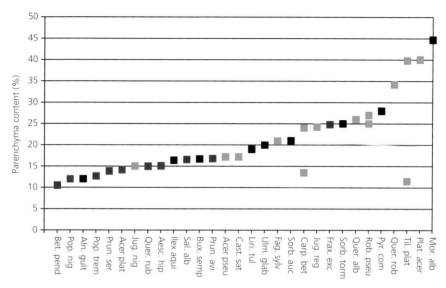

Figure 11.5 Parenchyma content of different tree species (Frey-Wyssling and Aeberli,1942; Wagenführ, 2007). Red: weak compartimentalizers. Green: effective compartimentalizers. Black: no classification published up to now.

Beyond this, quite different parenchyma contents have been published, for example for *Tilia platyphyllos* (Frey-Wyssling and Aeberli, 1942; Wagenführ, 2007). The outcome of this is a different classification of ability to compartmentalize (Baum and Schwarze, 2002).

The need for further research on content of parenchyma in wood of different tree species and its individual variation or age dependency is obvious. Other parameters of wood anatomy should also be recognized because, for example, very wide vessels in spring wood reduce the ability to compartmentalize (see Eckstein *et al.*, 1979). Nevertheless, the difference in parenchyma content of weak and effective compartimentalizers seems precise enough to classify tree species. Broad-leaved species with parenchyma content of more than 20% (for ring-porous trees, > 25%) may be regarded as effective compartimentalizers.

Conifers, in general, have a quite low content of parenchyma so, for conifers, this content is not useful as a criterion for classification. Compared to broad-leaved species, some other features, including low-diameter tracheids, resin channels, or substances such as taxol in yew (*Taxus baccata*) support compartmentalization. Hence, some genera like *Pinus* or *Taxus* are classified as effective compartimentalizers, while others like *Picea, Tsuga* or *Thuja* are considered to be weak.

Also, the *ability of trees to reiterate* is species-specific (see Chapter 7, section 7.3). Regarding pruning, the ability to reiterate is of differing importance:

1 For trees with a high ability to reiterate (that also form adaptive reiterations), it is quite easy to reduce the crown, or parts of it, without substantially changing its original and species-specific habit.

2 To a certain extent, the ability to reiterate provides some information on the intensity of reaction of the tree to a pruning procedure – the higher the ability, the more intensive the reaction. The point is the re-establishment of a balanced root/shoot relationship without unintended development of habit or branching patterns.

The *vitality* of trees is measured with regard to its vigorous growth and branching patterns (see Chapter 7). It is, therefore, a reliable sign for the amount of reserves within the system "tree" that are available for compartmentalization and wound closure. These reserves also influence the intensity of new shoot formation.

The influenceable parameters of pruning that are essential for its success are discussed subsequently.

11.4 The pruning system

The faster and the more effective a tree is able to compartmentalize a wound, and to prevent pathogens from penetrating into the trunk until it is completely encapsulated, is the major factor influencing the decision about the right way to prune.

Detailed research on the right pruning system was carried out at the end of the 1970s and in the 1980s by Shigo (1989) in North America and at the Institute of Arboriculture, Hamburg, Germany (Dujesiefken and Stobbe, 2002). The results confirmed the advantages of pruning at the edge of the branch collar. The branch-shedding collar is a collar-like diameter jump at the base of a branch, visible on dead (but also some living) branches and indicating the imminent death of the branch (Figure 11.6; and see Chapter 8, section 8.3). Thus, in Germany, it is literally translated as "farewell-collar" (Matthek and Bethge, 2014).

Further research has been done about wound reactions after pruning branches without a branch collar, or with included bark in the fork. From the results, the Hamburg Tree Pruning System was developed and, since 1992, has been integrated into the rules and regulations for tree care methods in Central Europe (FLL, 2006). It is described here.

Regarding the Hamburg Tree Pruning System in general, every branch must be removed twice, following a three-cut-method (see also ANSI, 2008; Figure 11.7). The first cut needs to be made on the underside, to the depth of about one third of the branch. The second cut is on top of the first cut (if using a chainsaw) or slightly beyond (if using a hand saw) (Gilman, 2012). Thus, a branch stub is produced that needs to be properly removed by the final cut. This method prevents cracking the wood during the removal of the branch, and tearing the bark, which would lead to unnecessary big wounds.

It is also fundamental to reduce a branch always to a lateral branch (supply branch). This ensures that the essential water flow, as well as the supply of assimilates to the axes and pruning surface, is maintained (Roloff, 2004). A heading cut is less desirable (Gilman, 2012).

Figure 11.6 Living (left) and dead (right) branch of *Tilia cordata* with branch-shedding collar.

Additionally, the production of several new sprouts at the head or below the cut is prevented, due to the hormonal control of the lateral branch. No branch stub may remain, because it is not supplied by assimilates, will die back, and will provide habitat for pathogens. The lateral branch should be no less than one-third the diameter of the cut trunk or branch (Figure 11.7), and it needs to be able to provide sufficient assimilates for compartmentalization and wound wood formation. This is difficult to ensure in the case of a very intense reduction, where another crown-securing system should be preferred to pruning.

Of course, it is always helpful that sharp and clean tools are used for pruning, to prevent dispensable damage by squeezing living tissue, leaving a rough pruning surface or allowing the unintended distribution of pathogens, for example *Fusarium* or *Verticillium* (Gilman, 2012).

Proper pruning of a branch with branch collar must be right outside the branch bark ridge, slanting downwards in accordance of the shape of the branch collar. The collar must remain at the trunk (Figure 11.8), and remaining tissue must not be damaged. Studies have shown that, in the area of the branch collar, a protection zone against

Figure 11.7 Principles: first prune to a stub, then remove the stub with a final cut (three-cut-method) (left). For a reducing cut, prune to a lateral branch with at least one-third of the diameter of the reduced trunk or branch (D: diameter).

Figure 11.8 Pruning to a branch collar.

Figure 11.9 Pruning without branch collar.

pathogens is pre-formed that may not be penetrated by the cut. A branch collar is a clear sign of effective compartmentalization at this point (Eisner *et al.*, 2002). To remove dead branches, it is also necessary to respect the branch collar.

In case there is no branch collar visible (Figure 11.9), according to the Hamburg Tree Pruning System, the cut has also be right outside the branch bark ridge, but straight down, to prevent cambial die-back due to a supply shadow. Other authors (Shigo, 1989; Gilman, 2012) advise to prune at a slight angle to the outside, reducing the wound size. In many cases, especially if the branch is ascending in a very steep angle, this seems more feasible.

Branches with included bark in the fork have to be pruned outside the lib-like rib or ridge, and the cut must be straight, without damaging the remaining trunk above the wound (Figure 11.10). In spite of proper pruning on top of the wound, often no wound wood is developed, because here the cambium is badly served with assimilates due to the included bark (supply shadow). When pruning too close to the trunk, huge wounds can be produced. However, it is not possible to cut the branch in a right angle, because the branch stub, which is not supplied by assimilates, will show cambial die-back. If the branch diameter is too large, a crown-securing system should be taken into consideration to prevent wounds that are too large.

Reducing a co-dominant stem (Figure 11.11) means loss of a considerable part of the crown and assimilating surface. It usually produces huge wounds, and preferably should have been done in an early stage at the nursery. Also, for compartmentalization, the relationship between the diameters of the removed and remaining stems are relevant (Eisner *et al.*, 2002).

If the diameter of co-dominant stems gets too big (weak compartimentalizers > 5 cm; effective compartimentalizers > 10 cm), the stem should be reduced, not removed. When removing such stems, the cut must be made close to the remaining stem outside the branch bark ridge. Also, in this case, a crown-securing system might be preferable.

11.4.1 Palms
Palms are of special interest as urban trees worldwide, as they follow completely different architectural models of tree growth, compared with all broad-leaved trees or conifers (Hallé *et al*, 1978), especially because of lacking a secondary diameter

Figure 11.10 Pruning with included bark.

Figure 11.11 Removing a codominant stem.

growth. Many species do not ramify at all, so the previously described pruning system is not useful.

In most cases, it is not necessary to prune palms (Broschat, 2014). They are only pruned for aesthetic purposes, or for to keep dead fronds from falling down, or when they are interfering with pedestrian passageways. It is of special importance not to over-prune palms. In this case, the reduction in canopy size results in reduced photosynthetic capacity, development of a smaller trunk diameter and a severe shortage of assimilates for the ongoing regeneration of the root system (Broschat, 2014; Gilman, 2012).

Figure 11.12 Pruning palms: removing fronds below the horizontal line (green) is acceptable. More (red line) is overpruning.

Only dead or severely marred leaves should be removed, that are hanging below an imaginary horizontal line drawn through the bottom of the crown (Figure 11.12). Removal of all fronds below a 45° angle above the horizontal line was practiced in former times, but today is considered to be overpruning and is not acceptable (ANSI, 2008). If living leaves must be removed from palms, the leaf base should remain at the trunk, in order not to injure it and create an entrance for pathogens (Gilman, 2012).

11.5 Intensity of pruning

The most common pruning procedures are crown maintenance and crown reduction. These are defined as follows (compare FLL, 2006):
- **Crown maintenance:** removal of dead, ill, broken, damaged, crossing and rubbing branches, preventing aberrations by pruning of predominantly twigs (1–3 cm diameter) or small branches (3–5 cm diameter), if necessary removing of branch stubs.

• **Crown reduction:** pruning procedure cutting branches up to 5–10 cm in diameter (for example, if its security against fracture, or its structural integrity, is at risk, or the crown is not sufficiently supplied).

It is clear that already, by definition, proper pruning only removes branches up to a diameter of 10 cm maximum. All other procedures that remove thicker branches are exceptions – special procedures (emergency steps), or non-professional work. Due to a limited expectation of life thereafter, felling and replacement of the tree should be also considered from an economic point of view. For weak compartimentalizers, even removing branches with more than 5 cm in diameter is considered to be problematic, because discoloration and decay extend far into the trunk. In any case, intensity of pruning should be as low as possible, especially if a crown-securing system has a greater likelihood of achieving the targeted structural integrity or security against fracture.

It is difficult to respect this rule in practice. The irreparable sins of former times (e.g., topping) and unexpected situations (e.g., damage by heavy storm or pathogens) sometimes necessitate the cutting of branches that are much more than 10 cm in diameter. In these cases, it is necessary to realize the inevitability of further operations, and even a reduced lifespan for the tree. On the other hand, it becomes clear how important it is to prune in time, as early as possible – this is an investment that pays.

Another important consequence of pruning is modified shoot growth. Usually, trees react to intensive pruning and a heavy disturbance of the root/shoot-relation with the formation of many fast growing shoots. On the other hand, there is only slight growth as a reaction to slight pruning. This depends also on the ability of the trees to reiterate, and their vitality. Too much growth might lead to unintended changes of habit (e.g., competing shoots, loss of the dominant leader), up to the unacceptable consequences of topping (sprouts weakly attached and more prone to failure).

One-time heavy pruning of vigorous trees to make up leeway in a short time leads to heavy and undesired reaction by trees, followed by further operations and costs. In this case, less intensively repeated pruning procedures are recommended. This is a little similar to "retrenchment pruning" (Fay, 2003; Lonsdale, 2013), mimicking natural retrenchment by reducing the height and spread of the crown in gradual stages. In the final retrenchment, cuts are unavoidably going to be large in order to achieve the intended crown reduction. Thus, it is considered as a special procedure, like those defined earlier (crown regeneration cut; crown securing cut) (FLL, 2006) and is useful in the field of habitat creation in mature or ancient trees. Other crown securing systems should be preferred for tree hazard management.

11.6 Date of pruning

The ability of trees to react to pruning also depends on the season, as described previously. Selecting a suitable date of pruning, therefore, seems to be another quite important parameter for pruning success. In general, the ability to actively compartmentalize a wound after pruning living branches and encapsulating pathogens by wound wood formation requires the activity of living tissue in wood, as well as the availability of assimilates or usability of reserves. Removal of dead branches is possible at any season.

Assimilates are only available within the vegetation period. In the temperate zone of the northern hemisphere, this is from May to September (Gruber *et al*, 2010; Kozlowski *et al.*, 1991). The amount of assimilates in living wood is highest at the end of the vegetation period (Larcher, 2003). Compartmentalization is considerably delayed when

Figure 11.13 Flowering Tabebuia in Brazil. In this period, without assimilating leaves and high energy demand for flowering and fruit formation, only very few reserves are available for compartmentalization.

pruning after or before the vegetation period (Dujesiefken *et al.*, 2005) because, at these times, living tissues are inactive and reserves not usable. Compartmentalization particularly requires reserves that have already been consumed after growth at the beginning of the vegetation period (Larcher, 2003). Therefore, intense pruning shortly before or at the very beginning of vegetation period is not recommended (Gilman, 2012; Lonsdale, 2013).

In other regions, a dormant period might not only be induced by differing temperatures in different seasons, but also by dry periods, flooding, or even special flowering seasons (Figure 11.13). It is crucial to select the date of pruning in such a way that the living tissue is active and is supplied as well as possible with water and nutrients. Avoid pruning when a tree is stressed.

Some other aspects need to be regarded when deciding on a suitable date of pruning. These are described below:

11.6.1 Reduction of assimilates and reserves

Pruning in the vegetation period reduces the assimilating leaf surface and, therefore, the amount of energy available to the tree. For some species, pruning at the very beginning of the vegetation period, within a phase of mobilizing reserves, leads to intense sap flow (so-called "bleeding" – e.g., birch, sugar or Norway maple, walnut). Water loss is low (see Chapter 3, section 3.2), but loss of energy can negatively affect compartmentalization in older less vigorous high-maintenance trees. On the other hand, this sap is rich in carbohydrates, and covers the wound surface, where it is colonized by mould fungi, antagonistic to wood-decaying fungi.

Sprouting and growing new shoots consume lot of assimilates, which are then not available for compartmentalization. Loss of energy, especially within this phase of mobilization and sprouting, is therefore regarded as critical for an effective compartmentalization. Additionally, bark and cambium are easily damaged and separated from wood when pruning takes place during a growth flush, because early xylem is gelatinous and slippery (Gilman, 2012).

For very fast-growing species, and trees with a high ability to reiterate, this energy loss might also be intended to reduce reaction and new growth after pruning. This is practiced in orcharding, with early spring pruning for fast-growing cultivars. Here, only young and vigorous trees, with low branch diameters, are treated, which compensates for the disadvantage of less effective compartmentalization in this season.

11.6.2 Species and nature conservation

The life cycle of many animals is closely linked to the vegetation period, because food supply is then high. The main breeding and nesting season is usually at the beginning of the vegetation period, so pruning is better executed at the end of this season. Of course, at any time, it is obligatory to respect the nests or habitats of rare, endangered and often protected animal species, no matter whether these are (for example) birds, bats or insects.

11.6.3 Hazard of fungal infestation

Spores of fungi are everywhere, and fungal infestation of the pruning wound may not be preventable. Of course, infestation pressure is enhanced in the vegetation period.

11.6.4 Risk of sunburn

Crown reduction may cause overheating of cambium by enhanced radiation. This applies especially for solitary thin- and dark-barked trees in summer. Laminar die-back of cambium leads to huge wounds and fungal decay. In the tropics, the risk of sunburn is present in any season, but not only the cambium is sunburned – shade-adapted leaves or needles may also be badly damaged, especially for evergreen species. In regions with high solar radiation and also high ultra-violet radiation, less intense and repeated pruning is recommended in order to ensure individual adaptation of the tree.

11.6.5 Severe frost

When pruning in winter, severe frost is also possible, and any laid-open living tissue might freeze. The importance of this tissue for compartmentalization and wound closure has already been highlighted. Its damage must be prevented.

11.6.6 Visibility

Dying or dead branches are more easily visible within the vegetation period, due to fewer or missing leaves. The structure of the whole crown is better assessed without leaves (FLL, 2006; Dujesiefken and Liese, 2008).

11.7 Conclusion

In selecting a suitable time for pruning, and appropriate intensity of pruning, following the Hamburg Tree Pruning System and regarding species-specific features and individual vitality, sufficient and practicable parameters for professional tree pruning are available.

References

ANSI (American National Standards Institute) (2008). *American National Standard A 300 Part 1 for Tree Care Operations – Tree, shrub and other woody plants maintenance – Standard Practices (Pruning)*. Washington, DC.

Barry, K., Pearce, R. and Mohammed, C. (2000). Properties of the reaction zones associated with decay from pruning wounds in plantation-grown *Eucalyptus nitens. Forest Pathology* **30**, 233–245.

Baum, S. and Schwarze, F. (2002). Large-leaved lime (*Tilia platyphyllos*) has a low ability to compartmentalize decay fungi via reaction zone formation. *New Phytologist* **154**, 881–490.

Broschat, T.K. (2014). *Pruning Palms*. ENH1182. UF/IFAS Extension. http://edis.ifas.ufl.edu/ep443 [accessed March 2015]

Dujesiefken, D. and Liese, W. (2008). *Das CODIT-Prinzip. Von den Bäumen lernen für eine fachgerechte Baumpflege*. Haymarket Media, Braunschweig.

Dujesiefken, D. and Stobbe, H. (2002). The Hamburg Tree Pruning System – A framework for pruning of individual trees. *Urban Forestry & Urban Greening* **1**, 75–82.

Dujesiefken, D., Liese, W., Shortle, W. and Minocha, R. (2005). Response of beech and oaks to wounds at different times of the year. *European Journal of Forest Research* **124**, 113–117.

Eckstein, D., Liese, W. and Shigo, A.L. (1979). Relationship of wood structure to compartmentalisation of discolored wood in hybrid poplar. *Canadian Journal of Forest Research* **9**, 205–210.

Eisner, N.J., Gilman, E.F. and Grabosky, J.C. (2002): Branch morphology impacts compartmentalization of pruning wounds. *Journal of Arboriculture* **28**, 99–105.

Fay, N. (2003). *Natural Fracture Pruning Techniques and Coronet Cuts*. http://www.arborecology.co.uk/resources/coronetcuts_naturalfracture.PDF [accessed March 2015]

FLL (Forschungsgesellschaft Landschaftsentwicklung Landschaftsbau e. V.) (2006). *Zusätzliche Technische Vertragsbedingungen und Richtlinien für Baumpflege – ZTV-Baumpflege*. Selbstverlag, Bonn.

Frey-Wyssling, A., Aeberli, H. (1942). Der Anteil von Fasern, Gefäßen und Parenchym verschiedener Holzarten in Dreiecksdarstellung. *Holz als Roh- und Werkstoff* **5**, 265–268.

Gilman, E.F. (2012). *An illustrated Guide to Pruning*, 3rd edition. Delmar. Clifton Park, NY.

Gruber, A., Strobl, S., Veit, B. and Oberhuber, W. (2010). Impact of drought on the temporal dynamics of wood formation in *Pinus sylvestris. Tree Physiology* **30**, 490–501.

Hallé, F., Oldeman, R.A.A. and Tomlinson, P.B. (1978). *Tropical Trees and Forests, an architectural analysis*. Springer. Berlin, Heidelberg, New York.

Kozlowski, T.T., Kramer, P.J. and Pallardy, S.G. (1991). *The Physiological Ecology of Woody Plants*. Academic Press, San Diego.

Larcher, W. (2003). *Physiological Plant Ecology*, 4th edition. Springer Verlag. Berlin/ Heidelberg.

Lonsdale, D. (ed, 2013). *Ancient and other veteran trees: further guidance on management*. The Tree Council, London.

Mattheck, C. and Bethge, K. (2014). *Die Körpersprache der Bäume: Enzyklopädie des Visual Treee Assessment*. KIT Karlsruhe.

Roloff, A. (2004). *Trees – Phenomena of adaptation and optimization*. Ecomed, Landsberg

Schwarze, F.W.M.R., Engels, J. and Mattheck, C. (1999). *Holzzersetzende Pilze in Bäumen*. Rombach Verlag, Freiburg.

Shigo, A.L. (1989). *Tree Pruning*. Shigo and Trees Associates. Durham.

Wagenführ, R. (2007). *Holzatlas*, 6th edition. Fachbuchverlag, Leipzig.

Watson, W.T. (2005). Influence of tree size on transplant establishment and growth. *Horticultural Technology* **15**, 118–122.

CHAPTER 12

Transplanting large trees

Ulrich Pietzarka

Technische Universität Dresden, Tharandt, Germany

12.1 Introduction

There seem to be no technical limits anymore for transplanting old and also large trees. Many well-documented examples from Australia (Figure 12.1), North America, Asia or Europe prove this to be so (Jim, 1995; Thomas, 2012; Eveleth, 2013; Siegert, 2013; Anderson, 2014; GHK, 2014). The biggest ever transplanted tree claims to be a 750 year Ginkgo (*Ginkgo biloba*) which was 29 m in height, had a 33 m crown diameter and a total weight of 1250 tons (Guinness World Records, 2015). Even in standard catalogues of nurseries, solitary trees up to a height of 12 m and trunk diameter up to 40 cm are offered.

What are the frame conditions that need to be ensured before starting such a cost-intensive project? This chapter will provide some basic information and decision-making support. Additionally the specific local, technological, organizational and financial possibilities need to be recognized in every single case.

12.2 Definitions, tasks, decisions

A large tree, according to a definition given in the regulations for transplantation of large trees in Central Europe (FLL, 2006), is a tree that has a trunk circumference of 30 cm already, measured one meter above the ground, which means a trunk diameter of just about 10 cm. Descriptions given here refer to trees that do not originate from production nurseries, but which have been established in landscapes for quite a long time. Such trees have not been adapted to this process by regular transplanting (FLL, 2006; Harris *et al.*, 2004), so details on the preparation of the tree (pre-transplanting care) are necessary. High-quality nursery trees usually meet these requirements. Besides consideration of some special techniques and procedures for transplanting a tree in any case, it is crucial to ensure adequate post-planting care. This takes several years, and usually much longer than for smaller trees. Any negligence in care causes, very often and very quickly, a considerable loss in quality – up to total loss and death of the tree – implying a complete economic fiasco.

The reasons for transplanting large trees are very often politically motivated. The demands of people to preserve beautiful, beloved, cultural or historic important, and sometimes rare, trees that need to be removed from their original site are central (Jim, 1995).

Urban Tree Management: For the Sustainable Development of Green Cities, First Edition. Edited by Andreas Roloff.
© 2016 John Wiley & Sons, Ltd. Published 2016 by John Wiley & Sons, Ltd.

Figure 12.1 Transplanting a 750 year old boab (*Adansonia gregorii*) in Australia (Photograph by Jeremy Thomas, Botanic Gardens and Parks Authority Western Australia. Reproduced with permission of the Botanic Gardens and Parks Authority Western Australia).

Occasionally, special climatic effects, or those in styling urban areas, are demanded at once, rather than after a few years of growth (Schröder, 2008).

Before transplanting, it is necessary to make a decision about such a costly and laborious operation, which should be based on a balancing consideration of many factors, as shown in Figure 12.2 (see GHK, 2014). The first question is whether there is adequate *time for preparation*. The earlier it can start, and the more comprehensive it is, the more bearable the transplanting procedure is for the tree (Schröder, 2008). Therefore, while planning a project, adequate time for preparation is demanded. The whole transplanting operation takes at least 12 months, apart from time for consultations or approvals (GHK, 2014). It also has to be clear whether a *suitable receptor site* is available. Here, the future development of the tree needs to be considered – a point neglected in so many examples. There must be sufficient root space, as well as aboveground space for growth, and distances to traffic infrastructure or buildings need to be maintained. Also, a stopover at another place in a transit nursery might be possible, if good care is maintained. Of course, costs for additional maintenance and transport will rise.

A very important question for making the decision concerns the *suitability* of this special tree for transplantation. In general, the tree should show good vitality and be free from

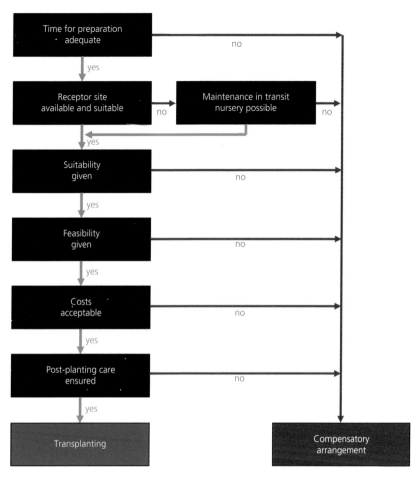

Figure 12.2 Decision-tree for transplanting. All requirements need to be fulfilled.

damage or aberrations. Otherwise, survival of the tree may be limited, or future problems of safety may follow. Trees in a stagnation or even resignation phase (see Chapter 7) are not suitable. Also, the life expectancy of the large tree needs to be long enough that a sufficient time for use is ensured.

The rareness of the tree species is regarded positive, while its potential to be invasive is negative. It might have a special historic or cultural importance. Not least, an urban tree needs to be attractive and good-looking – characteristics that are really difficult to measure. If excessive pruning operations are considered to be necessary for transplantation, which may ruin its natural form and are not professional (Jim, 1995), the tree is not suitable for transplantation. It must also be considered that development of a large tree at the receptor site is delayed compared to smaller trees (Watson, 1985; Watson, 2005; Gilman *et al.*, 2013), and also that mortality is much higher (Struve *et al.*, 2000; Struve, 2009; and see Chapter 6). Therefore, it should be questioned whether the less expensive planting of a smaller tree might have the same effect after a very few years.

If suitability of transplantation is given, then *feasibility* also needs to be evaluated. First, there are the technical resources available for transplanting. Today, the dimensions of the tree, especially its volume and weight, play a minor role. There are technical solutions; but is the technique useful without causing damage to the tree? Are recent and receptor sites also accessible; is the road between the sites and its load-bearing sufficient for the technique?

It is also very important that a reasonable root ball size can be achieved and transplanted, which is capable of sustaining and continuing the growth of the tree while balancing other logistical and cost concerns. The root system is developed individually, especially in urban sites. It does not follow special types of root system (see Chapter 4). Even if not expected, it might be shallow and very far-reaching. Other buildings or supply lines may cause difficulties in digging the root ball. Mapping out supply lines in combination with excavations to search for roots, or even soil radar, will help to evaluate the root system. Damage to roots with diameters of more than 5 cm must be prevented. If the root ball cannot be prepared such that damage is reduced to a few roots of 2–5 cm in diameter, then feasibility of transplantation is not a given.

The evaluation of all these described factors and parameters usually makes it necessary to order a professional consultant for a study of feasibility. Based on this study and the service specifications, an *estimation of costs* is possible. This estimate will be another major fact to be considered when making a decision about transplanting. It has to include costs for prolonged and intensive maintenance of the transplanted tree.

Transplantation of a large tree makes sense only if all these decisions are positive. The operations then necessary are described below.

12.3 Preparation

It is essential to design a transplantation operation around securing survival of the tree, stimulating its development and minimizing pruning and maintenance. These operations usually need to start at least one vegetation period before transplanting – better still, two (FLL, 2006).

The most important operation for preparation is digging a root trench. Normally, the root system is quiet widespread, especially if the tree originates from woodlands. Lifting such trees without initial preparation of a root ball will result in much of the root system being left in the soil (GHK, 2014). Therefore, it is necessary to dig a trench about 20 cm wide around the future root ball, down beneath the main root area (usually not more than 1.5 m). The distance between trench and trunk base must be calculated in such a way that the outer rim of the trench equates to the root ball size (see below).

It is advisable to dig the trench in stages that reflect segments of the circumference (stage digging) (FLL, 2006; GHK, 2014). Laid-open roots need to be pruned properly, using sharp tools. In case there are one or two thick roots (if there are more, feasibility is not given), these should be left for static reasons, or another securing system needs to be installed. The excavated trench needs to be back-filled with amended soil mix, with growth stimulants or fertilizers. Growth of new fibrous absorption roots from the cut zone has to be encouraged here before transplanting, to stimulate regeneration of the root system.

The trench frequently needs to be watered deeply (FLL, 2006). In a few cases it might be useful to aerate the root ball by pressurized air (Schröder, 2008). However, if the future root ball is already heavily affected by, for example, soil compaction or element inputs, then feasibility of transplanting is not given. Compass orientation of the trees should be marked, in order to plant it at the new site in the same orientation (ANSI, 2005). This is especially necessary for evergreen species, because their leaves are adapted to radiation.

12.4 Transplantation practices

The best season for transplantation is dormancy (ANSI, 2005; FLL, 2006), although this will vary between species (Etemadi *et al.*, 2013). Evapo-transpiration should be as low as possible (GHK, 2014).

The size of the root ball is of special importance. In general, it should be of sufficient diameter and depth to encompass enough of the root system as necessary for establishment (ANSI, 2005). Most publications demand a root ball size ten times the trunk diameter (which means, for a trunk 10 cm in diameter, a root ball diameter of one meter) (Schröder, 2008; FLL, 2006; GHK, 2014). American standards demand a little more, namely 30 cm root ball diameter per inch (2.54 cm) trunk diameter (i.e., for a trunk with 10 cm in diameter, a root ball diameter of 118 cm) (ANSI, 2005). If such a size is not possible to ensure, for instance if the root system is formed very asymmetrically or disturbed in any other way, which is often to be observed in cities, then feasibility for transplantation is not given.

For transplanting palms, the situation is different. Their root ball may be much smaller, because adventitious roots are formed constantly in a special zone near the trunk base (Harris *et al.*, 2004). The minimum root ball radius should be 15 cm (6 inches) from the base of the trunk at ground level, although root balls larger than this should be preferred. The root ball depth should be at least 1.5 times the root ball diameter (ANSI, 2005).

Recent machinery for transplanting large trees (Figure 12.3) ensures a complete, undamaged root ball while lifting, transport and planting is operated in one step. If

Figure 12.3 Tree spade mounted on a wheel loader (Photograph by A. Roloff).

another workflow is planned, or another technique for lifting and transport is available, the root ball must be secured and maintained. This may be carried out by boxes, frames or metal nets around the root ball (see Jim, 1995; Siegert, 2013). The root ball needs to be protected against dehydration, and needs to be watered. It is also advisable to water it before lifting and transport (GHK, 2014).

For transport, trees should not be lifted by the trunk (as this can cause serious trunk injury), but by the root ball, which should be properly prepared and wrapped. Trunk and crown may only be carefully secured to prevent any damage to bark and cambium by bruise or abrasion. Root balls that are not properly protected would easily collapse under their own weight during transplantation (GHK, 2014).

Often, trees need to be transported in a horizontal position. The root ball must not be damaged and must be properly wrapped below. The crown of the tree should also be carefully wrapped, to minimize the risks of dying, branch damage due to excessive movements, and wind damage or increased transpiration (GHK, 2014).

For planting at the receptor site, a planting pit of sufficient dimensions (at least 12 m³), and with appropriate soil, must have been prepared. Into this planting pit, the planting hole needs to be excavated. Using machinery for transplanting, the hole has the same dimension as the root ball, although the hole needs to be bigger if using another planting technique. A size of 1.5 times the root ball (ANSI, 2005), or root ball size plus 30 cm in any direction, is recommended (FLL, 2006). According to Watson and Himelick (1997), a planting hole that is two to three times the width of the root ball at the surface is optimum. The soil of the planting pit, and the substrate added to the planting hole, should support fast regeneration and growth of the root system beyond the planting hole. Therefore, it might be fertilized and provided with drainage to allow effective percolation of water (GHK, 2014).

Planting has to be carried out in a way such that the root ball is on the same level as before; the surface of the root ball must not be below or above the surrounding soil. This is important to secure a sufficient air supply for the roots (critical, in case of planting too deep and covering the root ball with soil) and to prevent dehydration of the root area (in case of planting too high). If a complete exchange of the soil in the planting pit is necessary, later settling must be prevented or considered, so a slight compaction of the filling material will be necessary beneath the root ball.

All root ball supporting materials should be removed from the planting hole prior to back-filling. The back-fill soil shall be reinstated and settled in layered sections, to limit future settling and to prevent air pockets at the side and beneath the root ball. Intensive watering right after planting guarantees a close contact between root ball and the surrounding soil. Around the former planting hole, a tiny wall of soil should be developed to prevent run-off of irrigation water. The area of the planting pit should be mulched in order to conserve soil moisture, to buffer soil temperature extremes and to control weeds (GHK, 2014).

Transplanting large trees always makes it obligatory to retain the tree in an upright position by support systems after planting. This needs to be effective for several years, and must prevent any damage to the trunk or root system. Different systems are in use. In general, support by staking or guying with ropes and ground anchorage is installed in three or four directions. The support system needs to be fixed to the tree right below the crown. It has to be properly padded to prevent any damage by bruise or abrasion.

After planting, a slight pruning operation as crown maintenance to remove minor damages, rubbing branches or other aberrations is helpful. Crown volume is a little reduced, and so adjusted to the root system. Painting the trunk white helps to prevent the cambium from sunburn and overheating. Radiation conditions should have been assessed already, while evaluating feasibility to transplant.

12.5 Post-planting care

The importance of post-planting care cannot be emphasized enough, as it is crucial for the success of the whole operation. Guaranteed maintenance of an appropriate standard and intensity, and for a long enough time to allow the tree to recover, is very important for the decision to transplant a large tree. Of course, it is not possible to give reliable and exact information on the duration, intensity, or operations necessary. These will depend on the individual site, species, size, local conditions, climate and many more factors. In general, maintenance takes much more than one year – up to five – and intensity depends on the age and size of the tree. Also, depending on climate conditions, 15–20 watering operations per year might be necessary, with the amount of water depending on the tree's dimensions. As a guiding value, approximately 200 liters for a tree with trunk diameter of 10 cm, or 1000 liters for a tree with trunk diameter more than 25 cm, are required (FLL, 2006). This is not done by a watering can.

Pest control or fertilization are required in case of heavy infestation or confirmed nutrient deficiency. Support systems need to be checked periodically, for as long as they are necessary. Supports that stay too long without proper adjustment will do more harm than good to the trees (GHK, 2014).

12.6 Conclusion

Taking into account all the described requirements, and based on a complex and well-balanced decision process, today there are technical solutions for transplanting large trees. In this way, old, valuable and beloved trees may be preserved, in case all other possibilities are exhausted to keep the tree in its place. Nevertheless, transplanting large trees is regarded as exceptional, because it is time-consuming, expensive, and it holds some risks that are difficult to calculate.

References

Anderson, N. (2014). *The Old Glory Story: A move rooted in controversy.* http://www.landscapeonline.com/research/article/5502. [accessed March 2015]

ANSI (American National Standards Institute) (2005). *American National Standard A 300 Part 6 for Tree Care Operations – Tree, shrub and other woody plants maintenance – Standard Practices (Transplanting).* Washington, DC.

Etemadi, N., Nezhad, R.M., Zamani, N. and Majidi, M.M. (2013). Effect of transplanting date and harvest method on growth and survival of three urban tree species in an arid climate. *Arboriculture & Urban Forestry* **39**, 211–217.

Eveleth, R. (2013). *It's crazy to move a hundred-year-old tree, but this one is thriving.* http://www.smithsonianmag.com/smart-news/its-crazy-to-move-a-hundred-year-old-tree-but-this-one-is-thriving-46475775. [accessed March 2015]

FLL (Forschungsgesellschaft Landschaftsentwicklung Landschaftsbau e.V.) (2006). *Zusätzliche technische Vertragsbedingungen und Richtlinien für das Verpflanzen von Großbäumen und Großsträuchern.* ZTV-Großbaumverpflanzung. Bonn.

GHK (Government of Hong Kong Special Administrative Region, Greening, Landscape and Tree Management Section, Development Bureau) (2014) *Guidelines on Tree Transplanting.* http://www.trees.gov.hk/filemanager/content/attachments/Guidelines_on_Tree_Transplanting.pdf [accessed March 2015]

Gilman, E.F., Miesbauer, J., Harchick, C. and Beeson, R.C. (2013). Impact of tree size and container volume at planting, mulch and irrigation on *Acer rubrum* L. growth and anchorage. *Arboriculture & Urban Forestry* **39**, 173–181.

Guinness World Records (2015). *Largest tree transplanted.* http://www.guinnessworldrecords.com/world-records/largest-tree-transplanted. [accessed March 2015]

Harris, R.W., Clark, J.R. and Matheny, N.P. (2004). *Arboriculture. Integrated management of landscape trees, shrubs and vines.* Prentice Hall. Upper Saddle River, NJ.

Jim, C.Y. (1995). Transplanting two champion specimens of mature Chinese Banyans. *Journal of Arboriculture* **21**, 289–259.

Schröder, K. (2008). Möglichkeiten und Grenzen der Großbaumverpflanzung – Dreißig Jahre Erfahrungen in Osnabrück. In: Dujesiefken, D. and Kockerbeck, P. (eds). *Jahrbuch der Baumpflege 2008.* Haymarked Media, Braunschweig.

Siegert, B. (2013). Zwei 150 t-Platanen im Schwebflug. *AFZ-DerWald* **68**, 36–37.

Struve, D.K. (2009). Tree establishment: A review of some of the factors affecting transplant survival and establishment. *Arboriculture & Urban Forestry* **35**, 10–13.

Struve, K.D., Burchfield, L. and Maupin, C. (2000). Survival and growth of transplanted large- and small-caliper Red Oaks. *Journal of Arboriculture* **26**, 162–169.

Thomas, J. (2012). A long road. *Landscope* **27**, 11–15.

Watson, G. (1985). Tree size affects root regeneration and top growth after transplanting. *Journal of Arboriculture* **11**, 37–40.

Watson, G.W. and Himelick, E.B. (1997). *Principle and Practice of Planting Trees and Shrubs.* International Society of Arboriculture, Savoy, Il.

Watson, W.T. (2005). Influence of tree size on transplant establishment and growth. *HorTechnology* **15**, 118–122.

CHAPTER 13

Dust and noise reduction

Britt Kniesel

Technische Universität Dresden, Tharandt, Germany

13.1 Dust

The attenuation of the amount of dust from air layers at ground level is especially important for citizens. Three factors are primarily responsible for reduction:

1 The physico-chemical characteristics of particles.
2 The micro-climatic conditions.
3 The ability of plants to filter out dust. (Slinn, 1982).

13.1.1 Dust definition and origins

Dust is defined as fine particles of matter, including smoke, soot and smog. It can be classified by its diameter: coarse particles (>10 μm); "fine dust" or particulate matter (<10 μm (PM10); < 2.5 μm (PM2.5)); and ultrafine particles (<0.1 μm UP). Coarse particles can still be filtered by the human nose, whereas PM10 can be inhaled, PM2.5 is already respirable and UP penetrates directly into the blood circulation. Dust can further be divided by its composition: organic (e.g., pollen, bacteria, spores, sawdust); inorganic (e.g., sand, sea salt, cement, metals); and mixed dust (e.g., house dust).

Sources and emitters of dust in urban areas are, for example: road and maritime traffic, aviation, dirt lanes, construction sites, industrial sources, firing installations, private housing (fireworks, BBQ), open waste incineration, and trees. External origins include forest fires, agriculture, volcanoes, desertification and open-cast mining.

13.1.2 Interaction between dust particles and vegetation

Dust should be filtered and sedimented at least temporarily. Wind plays a major role in dust transportation and sedimentation – the heavier and larger the particles, the faster they sediment. Wind conditions in urban areas can easily be controlled by reasonable plantations; vegetation can establish a flow resistance for wind, which becomes attenuated and deflected. This leads to relatively dust-free zones on the lee side of a plantation. Another benefit of roadside vegetation is the increase of ground and air humidity, which favors the passive sedimentation of dust. Similar effects can be achieved by uncut grass and herbs; it decreases temperature and wind, and increases humidity. In streets without trees, the dust impact is up to six times higher (Frankfurt, Germany, 1970s). Apart from that it should not be ignored that trees produce organic dust themselves, including pollen (Sofiev and Bergmann, 2013) and small fragments.

Urban Tree Management: For the Sustainable Development of Green Cities, First Edition. Edited by Andreas Roloff.
© 2016 John Wiley & Sons, Ltd. Published 2016 by John Wiley & Sons, Ltd.

13.1.2.1 Characteristics and strategies of plants

Plants do not "act" physically, but it cannot be ruled out that there might be a possible electrostatic attraction and fixation of dust particles. Due to the anatomical-morphological properties of leaf, twig and stem surfaces, plants have different abilities to filter or bind aerosols.

Characteristics which determine plants' *filtering performance* are mainly defined by crown structure and leaf positioning: leaf area, leaf angle, flexibility, the quantity of leaves per crown-volume and the duration of foliation. Plants with a stiff, uneven, dense and evergreen foliation are more effective in their ability to filter dust than plants with flexible, plane, sparse, and deciduous foliation. Conifers are more effective as filterers than are broad-leaved trees, due to the stiffness of their needles in wind, and their evergreen leaves. The *fixation of dust* is largely determined by the anatomical and morphological structures of leaves and bark – roughness, relief, hairs (living and dead), leaf venation, glands, wettability, concavities, and the morphology of the leaf edge (Beckett *et al.*, 2000).

Plants can be grouped into two strategists: dust accumulators, and dust attenuators. A proceeding deposition and accumulation of dust particles, which can only be rinsed or blown off partially, characterizes the first group. The second group benefits from the lotus-effect. Due to the surface structure, dust cannot find any hold, and is easily re-released by wind and rain (Thönnessen, 2002).

13.1.2.2 Case studies

In the first case study, Langner (2006) analyzed the filtration performance of a single roadside tree. A Norway maple (*Acer platanoides*), with a height of about 9 meters, had a total sum of leaves of 41,000. The averaged leaf area was around 68 cm², so the total leaf area was about 278 m². Throughout the vegetation period, the tree collected 2 kg of dust, whereof 20% was in the "fine dust" category (PM2.5–PM10). The *in situ* emissions of fine dust were around 3.5 kg. That makes a remarkable filtration performance by this single tree, of approximately 11% of the fine dust. However, due to the locally produced and dispersed fine dust, this calculation had to be adjusted to 4%. This study illustrates very clearly how restricted the filtration performance of a single tree is. Well-arranged plantations of trees combined with shrubs, formed into green belts, increase filtration and sedimentation substantially.

For the second case study, Thönnessen (2002) tested the effect of façade greenings on filtration of heavy metals. The study site was situated at a busy road (12,500 cars/day) in Düsseldorf, Germany. Leaves of the Japanese creeper (*Parthenocissus tricuspidata*) were sampled at five heights (2.0–13.5m), dried, and analyzed for the concentration of several heavy metals emitted by road traffic. The results provided a vertical and seasonal profile, with a maximum of heavy metal concentration at 2 m and a minimum at 4.5 m, with a general increase from spring to autumn (Figure 13.1).

The maximum is near the ground for two reasons – sedimentation and traffic-caused turbulences – which blow the dust into the façade greening. The minimum is also caused by turbulences, where the dust is transported out of the greening and down again. At higher levels, where the turbulences do not reach, passive sedimentation again dominates. Depending on the situation, especially in narrow street canyons, façade greenings are the best solution to improve air quality. Trees could even become counterproductive, as they disturb the interface to the fresh air from higher layers, and cause high concentration values underneath the crowns (Figure 13.2).

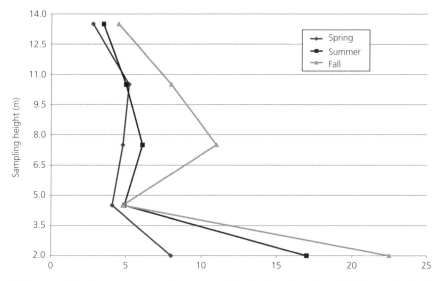

Figure 13.1 Generalized concentration of heavy metals in *Parthenocissus tricuspidata* leaves during the vegetation period (Adapted from Thönnessen 2002).

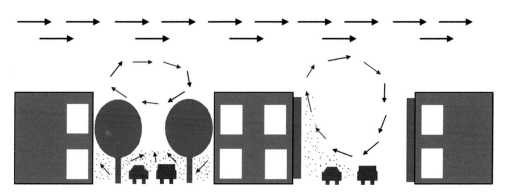

Figure 13.2 Street trees and facade greenings in narrow street canyons (arrows: wind and movement of dust particles, Adapted from Thönnessen 2002).

13.1.2.3 Effects of dust on vegetation

In addressing the demand for decreased dust in urban areas, it must not be neglected that dust might also damage plants. It can affect photosynthesis, respiration, transpiration, and may allow the penetration of phytotoxic gaseous pollutants. This might lead to visible injury symptoms, as well as decreased productivity (Farmer, 1993). A dust layer may disturb the light absorption: it increases in the infrared (750–1350 nm), and it decreases the reflection and transmission, inducing heat- and transpiration-stress. Leaves with a dust layer are distinctly warmer (2-6C°).

Dust can also interrupt the gas exchange of plants. Lenticels and stomata can be blocked, even in branches, twigs and stem. PM2.5 cannot intrude through stomata, but particles and aerosols < 1 μm. The damage is always dependent on many factors, including the distance to the emitter, the duration of the dust layer, the chemical composition of the dust and the meteorological conditions. In this context, the strategy to attenuate dust may probably be more advantageous over the year, whereas the dust accumulators partially damage themselves.

13.1.3 Planting design

The filtering ability of solitary plants is rather restricted, while vegetation stands are more efficient. Plantations should be near ground level, and as close to the source as possible, with green belts between road tracks and at the roadsides; the same applies for railway tracks. Plantations where all different characteristics are present are optimal – plants with stiff and uneven foliage, which would serve as filterers, in combination with plants which fix the dust on their leaf surfaces. The understory is important; the structure of a roadside plantation should be compact from the bottom up, beginning with grass and herbs (up to 0.75 m), followed by shrubs (2–3 m) and trees. The understory is necessary to catch the falling leaves and bind the dust to the ground to preserve a re-suspension (Figure 13.4).

13.2 Noise

Noise is defined as sound which is perceived as annoyance and disturbance. The main sound sources in urban areas are road and rail traffic, aircraft, industry, construction work and business. The traffic volume is increasing continuously, as well as people's expectation for quietness. There are numerous standards and regulations to impose noise control and limitations, but the costs of noise control have to be in balance with the environmental consequences. Most residential environments are affected by more than one noise source, and have various health effects, such as stress and sleep disorder.

13.2.1 Noise control

There are three steps required for noise level reduction:
1 Evaluation of the noise environment under existing conditions.
2 Determination of the acceptable noise level.
3 Determination of the difference between the two previous steps.
Techniques for noise control are related to control at the source, the receiver (less options for control) and the transmission path. At the noise source, it is possible to reduce the amplitude of vibrations, to reduce the motion of the components in vibration, and to use damping material. To control the transmission path, there are options to reduce the energy transmitted to the receiver and to introduce barriers of a larger size, compared to the wavelength of the noise.

13.2.2 Noise attenuation by vegetation

Solid barriers are rather effective in attenuating noise. Earth mounds, vertical screens, tunnels and walls achieve a reduction of around 20–40 dB(A). Similar to the dust filtration performance, the ability of plants to attenuate noise is restricted, especially for single plants. Solitary trees, narrow tree belts and hedges have virtually no protective function. In some cases, however, singular plants are effective, for example to break echoes between houses.

Generally, the attenuating effect highly depends on atmospheric conditions, including temperature, wind gradients and relative humidity. The maximal attenuation effect by vegetation is around 10 dB(A), whereas a reduction of 10 dB(A) results in a sensation of noise cut by half. An acceptable noise level for daytime activity is approximately 70 dB(A), and for evening time 50 dB(A). In order to have a similar effect to a 2 m high protective wall, tree belts of 25–30 m width are necessary. The most important effect, however, is optical noise attenuation. Plants reduce the individual noise sensation. If the noise source is invisible and, at the same time, hidden behind a relaxing green wall, the psychological effect is especially positive. This should not be underestimated.

13.2.2.1 Physical aspects

An acoustic wave hits different objects and surfaces and, depending on their features, it is absorbed, dispersed, reflected or refracted. Stem, branches and leaves have different effects. The smaller the objects (e.g., leaves), the smaller the refraction and scattering of waves; however, if the objects are numerous, the attenuating effect increases. Also, the bigger the scatterer, the lower the frequency at which the scattering effect occurs. The frequency itself determines the behavior of the acoustic wave: mid-frequencies (1 kHz) are mainly scattered, whereas high frequencies are likely to be absorbed. Following this, it can be expected that coniferous trees are effective in attenuating frequencies of 500–2000 Hz, and broad-leaved trees at 2000–8000 Hz (Bucur, 2006). This is a benefit, because loss of hearing is mainly caused by long-time impacts of higher frequencies.

13.2.2.2 Case studies

In the first case study, Tyagi *et al.* (2006) compared three different vegetation sites (belts of mixed trees, shrubs and grasses) and a control site (plain barren land) in Delhi (India) for traffic noise attenuation. They found that, in general, attenuation increases with frequency. Attenuation maxima can be observed at 400 Hz, 3.15 KHz and between 10–12.5 KHz. At the low frequency (400 Hz), attenuation is higher at vegetation sites, perhaps due to the "ground effect". Softer, more porous soil surfaces attenuate more at low frequencies. The second maximum, at 3.15 KHz, may be due to geometrical scattering of sound by stems and branches. A relative attenuation of more than 24 dB is observed at this frequency. The attenuation at the higher frequencies could be attributed to the foliage. The authors indicate vegetation belts as effective noise barriers along roadsides.

In the second case study, Fang and Ling (2003) evaluated the noise-reducing effect of four defined parameters (visibility, width, height and length) of vegetation belts. For this investigation, the authors analyzed 35 evergreen-tree belts of single species in Taiwan. A noise source was positioned in front of the belts, and the noise level was measured at different points within the belt. The results revealed a negative logarithmic relationship between visibility and attenuation, and a positive logarithmic relationship between width, length and height of the belt and attenuation. The lower the visibility, the higher the density; and the more branches and foliage reduce sound energy, the greater is the scattering effect. Visibility is the most influencing parameter, followed by width.

13.2.3 Planting design

The attenuation effect is dependent on species, age, crown structure and density. The bigger the leaves, the higher the leaf area index, and the longer the foliage is remaining, the higher is the noise attenuation. A tree belt may provide some immediate improvement in noise abatement, with the likelihood of better conditions as the plants mature.

As general recommendations, noise barriers should be as close to the noise source as possible. A green belt should be composed of different species of shrubs and trees, with

Figure 13.3 Schematic design of an optimal tree belt for attenuating traffic noise.

Figure 13.4 Optimal plantation for traffic noise protection in Copenhagen, Denmark (Photograph by A. Roloff).

(a)   (b)

Figure 13.5 a) Wooden fence within a plantation, foliage keeping species *Quercus robur*. **b**) lawn tracks
for noise reduction and dust sedimentation in the city of Dresden (Photographs by M. Meyer).

foliage from the ground up (Figures 13.3 and 13.4). The plants should be as close together
as the species will allow. A high diversity of species, with a range of foliage shapes and
sizes within the noise buffer, may also improve noise reduction (Nilsson *et al.*, 2015).
Generally, the growth requirements concerning air, water, nutrients, radiation, and space
for rooting have to be considered, as well as the final age, height, diameter, the living
crown ratio, and susceptibility to pests.

13.3 Conclusions

The planting designs for dust and noise reduction are very similar. Therefore, both aims
can be achieved by the same strategy. Combinations of solid barriers with dust and noise
reducing plantations (Figure 13.5a) are also reasonable. Specialties include lawn tracks
(Figure 13.5b), which combine the benefits of noise reduction with dust sedimentation,
by increasing humidity and bringing the "ground effect" into urban spaces at very small
dimensions.

References

Beckett, K.P., Freer-Smith, P. and Taylor, G. (2000). Effective tree species for local air-quality management.
    Journal of Arboriculture **26**, 12–19.
Bucur, V. (2006). *Urban forest acoustics*. Springer, Berlin, Heidelberg.
Fang, C.F. and Ling, D.L. (2003). Investigation of the noise reduction provided by tree belts. *Landscape and
    Urban Planning* **63**, 187–195.
Farmer, A.M. (1993). The effects of dust on vegetation – a review. *Environmental Pollution* **79**, 63–75.
Langner, M. (2006). Natürliche Filter? Die Filterung von Feinstäuben durch Stadtbäume. *Taspo* **3**, 20–23.

Nilsson, M.E., Bengtsson, J. and Klaeboe, R. (eds, 2015). *Environmental methods for transport noise reduction.* CRC Press Taylor and Francis, Boca Raton/USA.

Slinn, W.G.N. (1982). Predictions for particle deposition to vegetative canopies. *Atmospheric Environment* **16**, 1785–1794.

Sofiev, M. and Bergmann, K.-C. (eds, 2013). *Allergenic Pollen – A Review of the Production, Release, Distribution and Health Impacts.* Springer, Dordrecht.

Thönnessen, M. (2002). *Elementdynamik in fassadenbegrünendem Wilden Wein (Parthenocissus tricuspidata). Nährelemente, Anorganische Schadstoffe, Platin-Gruppen-Elemente, Filterleistung, Immissionshistorische Aspekte, Methodische Neu- und Weiterentwicklungen.* University Cologne/ Germany.

Tyagi, V., Kumar, K. and Jain, V.K. (2006). A study of the spectral characteristics of traffic noise attenuation by vegetation belts in Delhi. *Applied Acoustics* **67**, 926–935.

CHAPTER 14

Invasive species, indigenous vs. alien dendroflora

Matthias Meyer

Technische Universität Dresden, Tharandt, Germany

14.1 Introduction

Facing the fact of increasing global urbanization of habitats (Elmqvist *et al.*, 2013), contemporary urban tree management has to balance contrasting priorities of sustainability. On the one hand, the demand for perfectly functioning, cheap and healthy urban green has to be met with the lowest possible financial effort. Predicted extremes of climate conditions require making efforts to find plant material that is robust enough. On the other hand, though, the global decline of natural habitats corresponding to urbanization requires that the urban green sector becomes fully aware of the own responsibility for sustainability in gardening, and for preserving the (regional) natural heritage for future generations.

The Convention on Biological Diversity (CBD) pointed out that the introduction of invasive species is a major global threat to ecosystems. Invasions and urban biodiversity losses can have enormous impacts (Mooney, 2001; de Oliveira *et al.*, 2011). Also, there are plenty of possible motives for the urban landscaping or arboriculture sectors to put criteria of sustainability (e.g., the preservation of indigenous woody plant material) into practice. It is worth knowing basic floristic terminology and the risks of using alien material for the resident environment and citizenship, as well as knowing "first aid" measures in case of spontaneous impact. Implementing a few simple preventive measures would help to reduce risk and future costs. In some dramatic cases of invasive weeds, decisions about cost-intensive measures will be need to be made.

14.2 Floristic statuses – important definitions for urban dendroflora

In order to assess the selection between indigenous or "new" plant material in arboriculture or urban woodlands, it is essential to explain what is basically meant by, and how to differentiate between, different floristic statuses such as "indigenous" or "invasive". Dendroflora is a continuum, in that its specific traits and interactions are subject to everlasting environmental change and varying genetic constitution. There are transitional floristic stages that do not allow clear assignment and, furthermore, the determination of a certain species' floristic status depends on the definition of the related reference area (biogeographical range or political territory). The status can change with time.

Urban Tree Management: For the Sustainable Development of Green Cities, First Edition. Edited by Andreas Roloff.
© 2016 John Wiley & Sons, Ltd. Published 2016 by John Wiley & Sons, Ltd.

According to Pyšek *et al.* (2004), there are three floristic statuses: (I) origin status; (II) residence status; and (III) invasion (naturalization) status. These will be explained below. This classification is done mainly on the basis of reviews containing comparisons of, or suggestions on, standardized terminology on the floristic status of a given taxon (Schmidt and Wilhelm, 1995; Richardson *et al.*, 2000; Pyšek *et al.*, 2004; Kowarik and Rabitsch, 2010; Kowarik and Pyšek, 2012). A more academic classification system for statuses of indigeneity and naturalization, using relatively complicated Greek terminology, is presented by other authors (e.g., Sukopp, 1998).

14.2.1 "Indigenous" vs. "alien"

The terms *indigenous* and *alien* describe contrasting floral statuses of origin within the continuum of indigeneity. In the present text, any kind of plant material is denoted "indigenous" to an urban biotope if this biotope is located within the natural range of that plant material. The plant material has either originated within that territory without human involvement, or it has arrived this territory without intentional or unintentional human intervention (i.e., by evolutionarily expanding its native range; Pyšek *et al.*, 2004).

Plant material considered here can refer to a taxonomic unit (species, subspecies, local race, etc.) or to (remnant) stands of the previous or urban forests. In terms of arboriculture, it can also refer to commodities such as ornamentals, nursery stock or propagules (seeds, cuttings, etc.). The most important synonym to "indigenous" that is supported by leading ecologists is "native". However, in the present text, the term "indigenous" is preferred, in order to avoid confusion with dendroflora that was introduced a long time ago, has become naturalized and is now considered by some authors to be (new) native flora.

Another aspect is natural interspecific hybridization that can occur in some genera (e.g., *Eucalyptus, Salicaceae*). Interspecific hybrid material in its native hybrid zone is "indigenous" to urban habitats within that hybrid zone as well.

Another term that must be mentioned here is *autochthonous*. This term has similar definitions, but is not synonymous with "indigenous". Some European authors have stated that it should not be used to describe the indigeneity of taxa (e.g., Schmidt and Wilhelm, 1995). In forest science, restoration ecology, and by nurserymen, "autochthonous" is only used to qualify a forest stand, tree or shrub population, or a forest seed source of a certain provenance. A tree/shrub stand or population of an "indigenous" species is "autochthonous" if it has been continuously surviving in its respective (regional or local) territory with the help of natural rejuvenation for many generations since its immigration or speciation (Kleinschmit *et al.*, 2004).

Subject to natural selection, populations of important forest tree species have, over millennia, adapted to a wide range of varying geological or climatic site conditions. This has resulted in different ecotypes or local races (= provenances) that can only sustain full ecosystem services in areas to which they are adapted. Many forest authorities have developed a system of certification of "autochthonous" seed provenances, in order to guarantee supply with healthy and well site-adapted planting material. In many European countries, "autochthonous" planting material is also of rising importance in arboriculture and landscape arts (van der Mijnsbrugge *et al.*, 2010). The use of "autochthonous" provenances can be mandatory, for instance, in ecosystem restoration or remediation.

The opposite of the status "indigenous" is termed *alien* plant material here. Synonyms include "introduced", "exotic", "non-indigenous" or "non-native". Plant material is "alien" to an urban biotope if it is present "due to intentional or unintentional human involvement", or if it has established in the urban biotope without anthropogenic influence, but has originated from a habitat where it is "alien" (Pyšek *et al.*, 2004). Interspecific hybrid plant material is also "alien" to the respective urban habitat if it descends by crossbreeding, or spontaneously from an "indigenous" parent mated with an "alien" parent.

The *residence status* (Pyšek *et al.*, 2004), or the introduction date or time of first floristic evidence, can be valuable information for decision-makers planning the use of "alien" plant material in urban green. Recently introduced tree or shrub material that actually does not tend to become an urban weed can still become harmful later. This fact is due to considerable time lags in colonization (Petit *et al.*, 2004). "Alien" species introduced millennia ago do, most likely, represent a lower risk.

For European vegetation ecologists, it is relatively easy to define the initiation date of a globalised species exchange. *Archaeophytes* were introduced before modern times – that is, before the year of 1492 (≈1500 AD), when Christopher Columbus reached the coasts of the American continent. Archaeophyte tree or shrub species were introduced, for instance, by the Romans or the ancient Greeks from Central Asia or Persia. For example, due to their valuable fruits, the walnut tree (*Juglans regia*) and the peach tree (*Prunus persica*) were introduced to southern Europe, and have been spread over Europe in subsequent millennia (Kowarik and Rabitsch, 2010).

From 1492 on, oceanic and continental barriers were lifted very efficiently by globalised trade and travelling (Elton, 1958). Taxa introduced after 1492 are termed *neophytes*. Many "neophyte" tree species were introduced from North America, with a maximum intensity in the 18th century. With a later maximum in the 19th century, many species were introduced to Europe from East Asia or the Caucasus (Kowarik and Rabitsch, 2010).

14.2.2 "Casual" and "naturalized" vs. "invasive"

In vegetation ecology, the terms of any floristic status of naturalization are used neutrally, without connotations of any anthropocentric impact (Pyšek *et al.*, 2004). The most contrasting terms on a continuous scale of naturalization/invasion are *casual* and *invasive*. Colautti and MacIsaac (2004) have published an overview over classification models. In the present text, the barrier model of Richardson *et al.* (2000) is supported, which explains in chronological order how aliens become invasive by overcoming different barriers.

Alien plant material is "casual" to an urban habitat if it has overcome the introduction barrier. "Casual" alien plants are not, or only occasionally, able to reproduce at the place of cultivation. Furthermore, they do not form self-replacing populations by root sprouting. In contrast, alien plant material is able to become *naturalized* in an urban area if there are environmental conditions that allow the trees or shrubs to establish self-replacing populations that sustain for at least ten years with the help of auto-vegetative and/or sexual reproduction, but without human assistance (cf. Pyšek *et al.*, 2004). If a "naturalized" population of a certain woody species is able to overcome the barriers against spreading by the establishment of new (sub-) populations distant from the place of introduction, the taxon has become "invasive".

It must be noted that, in terms of vegetation ecology, only part of the "invasive" species can also become weedy or harmful for humans. Therefore, invasive woody plants that have an additional non-ecological but socio-economic impact should be termed "alien weed" or "invasive weed". In contrast, the anthropocentric view of normative regulations, international conventions or legal instructions does often combine the status of invasiveness in definitions with a socio-economic, health or environmental impact.

In the following sections, the focus is predominantly on alien species that have any kind of impact, namely invasive weeds. The ecological levels of organization which the invasion has an impact on can be distinguished into:
1 individuals;
2 populations; or
3 whole communities/ecosystems.
A small proportion of invasive species that have great ecological impact by reshaping nature of whole ecosystems are named "transformers". An example is the disappearance of nitrogen-poor grass, shrub or heathland as a consequence of invasions by nitrogen-fixing species (e.g., *Robinia pseudoacacia* in some parts of Europe). Due to their size, tree or shrub species easily have an impact on whole ecosystems (Hejda *et al.*, 2009). Reasonable classification systematics of ecological impact mechanisms, such as displacement of indigenous vegetation or genetic introgression, was recently compiled (Blackburn *et al.*, 2014; Kumschick *et al.*, 2015).

14.3 Invasibility of habitats and invasiveness of dendroflora in urban landscapes

Due to a high density of human-made habitats that are often the place of the first naturalization of an alien (Kühn *et al.*, 2004), urban areas play an important role in the first introduction phases. The invasibility of a certain urban habitat, and the invasiveness of an alien woody plant, should be assessed on a situation-specific basis. Depending on different site conditions, the efficacy of invasion barriers for the respective woody species is largely variable (cf. barrier model; Richardson *et al.*, 2000).

In each invasion state, different factors affect the barriers against an invasive plant (Pyšek *et al.*, 2009b). The first phase, the introduction itself, is mainly dependent on human activity (McGregor *et al.*, 2012; Rejmánek *et al.*, 2012). The bigger a city becomes, the more invasible it becomes (Pyšek, 1998). This is, for instance, due to higher agglomeration of introduction "gateways" like maritime or inland ports, railway and road traffic destination points. Also, grey- or brownfield areas, and many artificial or disturbed sites (Figure 14.1) without robust native plant communities (Rejmánek *et al.*, 2012) account for the high invasibility of urban rooms. The transportation system is heavily involved in tree seed dispersal inside urban areas (von der Lippe and Kowarik, 2012).

City areas are diverse in buildings, fences, kerb stones and other structures that provide edge zones, allowing wind-dispersed seeds to settle down (Figure 14.1). At the same sites, rain water, channeled from roads and buildings, as well as deposits of dust or organic matter (nutrients), are available. In combination with the relatively mild temperatures within city zones, this forms the basis for fairly invasible seedling beds for species that can later cope also with drought.

Very important for "alien" tree or shrub introduction are amateur gardeners, as well as the publicly organized landscaping sector. Hope *et al.* (2003) showed that family income is a significant predictor for urban biodiversity, indicating the influence of gardening activities in the private yard. Watering, fertilizing and the removal or suppression of indigenous weeds, subsumed as gardening activity, in some cases not only promotes the target species, but also promotes invasion by aliens.

During later phases, the invasion success is supported by species-specific traits that promote spread (Pyšek *et al.*, 2009a), rather than by taxonomic genus affiliation. However, some genera comprise species groups that can become alien tree weeds in many countries

(a)   (b)

Figure 14.1 a) Example from temperate climate zone of Central Europe (Dresden, Germany); a highly invasible city area comprising open land, boundary with international railroad, surrounded by urban greyfield sites partially covered with fertile invasive tree species that provide high propagule pressure (*Acer negundo, Ailanthus altissima, Robinia pseudoacacia, Populus* spp.). Countermeasure: planting native pioneer species (*Betula pendula*). **b**) Crannies for alien seedlings, *Ailanthus altissima* and *Populus* spp. (Berlin, Germany). (Photographs by M. Meyer)

(e.g., *Acacia, Acer, Pinus, Populus*). Once introduced into a foreign urban biotope, the success of a given alien tree or shrub in expanding the invasion area mainly depends on propagule pressure and specific traits that boost seed dispersal.

An ecological reproduction strategy (*r*-strategists, pioneer species), shorter lifecycles with early maturation; early and rich fructification; and lightweight seeds (wind dispersal) account for higher invasiveness (Carrillo-Gavilán and Vilà, 2010). Also, longer and/or earlier flowering seasons, and the ability to attract fruit vectors, such as birds, that will excrete the seeds at distant places, can promote spread. Among others, many species of the *Rosaceae* show this kind of vector co-adaptation.

Specific traits that promote resilience, including allelopathy, seed longevity, expansive root systems, intense production of root suckers, and the ability for auto-vegetative propagation from passively forwarded shoots or storage organs, as well as the ability to re-sprout very intensively after coppicing, are of minor importance for the mechanisms of propagule pressure. However, they account very much for weediness and habitat transformation.

14.4 Arguments pro or contra "alien" woody species and risk assessment

Some of the early purposes that justified woody alien introduction or intercontinental exchange are still very important in a global perspective: ornamental and forest tree breeding, fruit cropping, research and *ex situ* conservation, or simply having beautiful exotic trees. It seems that not only the use of introduced alien woods, but also *de novo* introductions, can be justified in some cases. Facing climatic change or anthropogenically altered site conditions, as well as the dramatic homogenization of anthropogenic dendroflora, foresters and nurserymen in many parts of the world attempt to find alien taxa that are better adapted to the predicted climatic conditions. This is especially true in the context of urban arboriculture that has to cope with even more extreme conditions (Roloff *et al.*, 2009).

In some cases, alien species perform better than their indigenous counterparts. This can partially be explained by the lack of antagonists or diseases (enemy release theory – e.g., Petit *et al.*, 2004). Often, the alien species can be seen as a useful complementation of relatively non-diverse subsets of indigenous city tree or shrub species. To summarize, there is a deeply complex background of arguments "pro" using alien material (see correspondence of Simberloff *et al.* (2011) with colleagues).

However, the same traits that make aliens desirable for urban tree management may partially result in weediness or in a socio-economic threat. This is the most important "contra" argument that decision-makers should keep in mind if they plan the use of woody aliens that are potentially invasive. The indigenous tree and shrub community is also a better choice for providing a full ecosystem function, for instance to the associated indigenous fauna.

When planning urban future stands of trees using aliens, it is also worth thinking about possible socio-economic problems that may result from weedy alien species. It must be noted, again, that aliens do not necessarily become invasive or weedy. However, introducing unknown invasive weeds can become a bad surprise, because invasive weeds can perform even better than at their origin (Parker *et al.*, 2013) or, respectively, better than indigenous weeds.

According to the enemy release hypothesis, the invasive weed is "unknown" to the competing indigenous vegetation, as well as to possible indigenous predators and diseases. For example, types of socio-economic impact or nuisance in urban rooms were illustrated by Kowarik (2011) and Chace (2013). Most prominent is the explosion of costs for maintenance of urban green and for the removal of tree weeds from railroads or roadsides, the release of new allergenic pollen, and displacement, adulteration and homogenization of resident urban green, as well as the release of volatile or poisonous organic compounds.

14.5 The example of the tree of heaven (*Ailanthus altissima*)

Ailanthus altissima, with nicknames "tree-of-hell" or "ghetto palm", is *the* example of an urban weed tree species in temperate and meridional vegetation zones on all continents. In non-urban secondary habitats, it is an invasive tree that can transform the flora, leading to decreasing diversity. This species came up with several eco-physiological adaptations that make it perform better in many cities than outside in the field, or in its natural habitats in eastern China. A map of its primary and secondary range was published by Kowarik and Säumel (2007), who also described the advantages and the nuisance deriving from its use.

After it was intentionally introduced as an ornamental to Europe and North America during the 18th century, it quickly became familiar with urban conditions there. Adapted to grow under water deficits in summer-warm habitats, and also on rocks and in disturbed habitats, it is very resistant to our cities' heat islands, dust and de-icing salt. This species follows an eco-physiological reproduction strategy combining extremely fast juvenile growth and short lifecycles (maximum 130 years) with early fructification (Figure 14.2). The terminal shoot growth continues throughout the whole growing period. Even though the seeds are not very lightweight, they occur abundantly and are very well transported in cities by wind and by all kinds of vehicles, especially along highways and railroads.

This tree is not easy to control, as well as costly to do so. It is able to spread by pronounced shallow root growth, combined with a taproot as storage organ. When affected by different above-soil stresses, such as chopping, coppicing or fire, tree-of-heaven responds with re-sprouting from stumps, and with the formation of root sucker thickets

Figure 14.2 a) *Ailanthus altissima* expanding stand probably from ornamental use, semi-arid near-desert climate (Columbia Basin, Boardman, Oregon, USA). **b**) One-year old sprout thicket (≈1.5–2.0 m) around girdled trees (cool temperate climate zone, Berlin). **c**) City tree in Litoměřice (cool temperate zone, Czech Republic). **d**) Planted tree row and beginning seedling invasion (summer-hot, dry climate, Okanagan Lake, West Kelowna, British Columbia, Canada). (Photographs by M. Meyer)

(Figure 14.2). The sprouts become as tall as approximately 2 m per year, with the maximum measured annual shoot length being around 5 m. Maintenance along railroads in German cities causes high costs for vegetation control, because the hazard zone of overhead wires can be reached very quickly.

14.6 Prevention and control measures against invasive woody species or tree weeds

According to Kowarik and Rabitsch (2010), types of measures that help to avoid problems with invasive species or invasive weeds can be divided into three groups, which correspond to the general chronological order of introduction phases. First, preventive measures can help to avoid the introduction, or to provide a basis of knowledge. Many countries in the world enclose rapidly growing urban zones, requiring plans for new urban forests and city trees. Planning this urban green, including the utilization of remnant forests as well as rather unknown alien species, makes it worth assessing the possible risk of invasion impacts. A helpful schematic of a decision tree is given in Ross and Lembi (2009) for the USA that could probably be adapted to other countries. Another decision tree was sketched by Kumschick *et al.* (2015), to help in structuring an adequate information search about possible impacts of an alien species.

The first stage of any assessment should be a database search for species or for experts. Richardson and Rejmanek (2011) have compiled a global list of woody invasive species, consisting of more than 600 taxa, including the weedy pests (comes with an *Excel* file supplement). The Global Invasive Species Database (flora and fauna) is linked at the website of the Invasive Species Specialist Group (ISSG.org, part of the International Union for Conservation of Nature, IUCN). The Global Invasive Species Information Network (GISIN.org) maintains a compiled list of online information sources and databases.

Decision-makers from Europe can search the DAISIE (europe-aliens.org; DAISIE, 2009) or the NOBANIS databases (nobanis.org) for a certain species. For more detailed information, control measurement instructions or other legal instructions, contacting local or governmental authorities is strongly suggested. Governmental authorities of some countries, or of states within larger countries, have compiled blacklists that can be mandatory for specific stakeholders, such as the nursery sector, official authorities in urban/roadside green, forestry, and so on.

If possible, genetic introgression pressure by potentially hybridizing species should be avoided, either by using sterile plants or by using indigenous alternatives that are genetically diverse, or which stem from autochthonous seed sources. From dioecious indigenous tree species, male individuals should be preferred, in order to avoid mating of indigenous females with alien pollen from surrounding alien males. Other possible preventive measures that can increase the invasion barriers of an urban habitat focus on early detection, detailed information activities for the local citizenship (e.g., city-owned website, local NGO partners, posters in garden centers, etc.), quarantine measures, monitoring and cooperation with research facilities.

Existing urban tree plantations can be inspected in the course of visual tree assessments for possibly critical situations indicating high invasibility, such as large construction activities, earthworks, or closed industrial zones close to stands of invasives. City governments could plan their need of financial and other resources in advance, according to an urgency ranking of adjacent invasive trees. If possible, the use of potentially seed- or root-contaminated soil substrates, as well as the use of potentially invasive or alien and non-diverse plant material, should be reduced. Best practice guidelines, self-commitment programs of the nursery sector, or compilations of indigenous alternatives seem to be possible instruments that could be enforced by regional authorities.

If it is too late for preventive measures, invasion management would be the next step. Generally it is best to start this as early as possible, with the maximum possible effort. However, it does not make sense to eradicate a certain tree collective, if the neighborhood is densely populated with other invasive trees providing intensive seed rain. Kowarik and Rabitsch (2010) published a decision matrix given to assess the suitability of countermeasures.

The effect of any countermeasure largely depends on the propagule pressure from the surroundings, and on the thoroughness of the workers. In need of particularly urgent action before seed release, fertile trees can be cut off first. As a very first and easy measure, trees could be girdled or treated with herbicide infusions. However, some tree species' invasions are promoted by frequent coppicing (Radtke *et al.*, 2013) or girdling. In small affected sites, an area that is large enough to cover stumps and the whole root system can be covered with opaque tarpaulins and sand/soil to avoid re-sprouting.

In some tree weed invasion scenarios, the application of herbicides or of specific antagonists is a possible measure aimed at reducing the invasion speed or eradicating dense woody stands (Ross and Lembi, 2009; Chace, 2013). However, both options require specific knowledge as well as absolutely species-specific application, and the assessment of environmental and health risks – otherwise, the resident indigenous vegetation or the human population could also be affected. If the defenses of the indigenous vegetation are impaired, this can stimulate an advance for the invasive weed that is often more resistant or vigorous to herbicides (Genovesi, 2011).

If nothing helps, and the invasion into the city becomes out of control, there is still a third option of action – acceptance of the invasion into the city, as well as developing ideas to draw benefit from the trees or shrubs as an indirect countermeasure. The trees could probably be "overused" as fuel wood in municipal energy supply plants. In some cases, shading the second generation by other trees can be successful.

14.7 Conclusions

The use of invasive tree or shrub species must be reduced. However, urban arboriculture seems to have more opportunities for the use of alien material without violating the criteria of sustainability, compared with rural landscaping and the forestry sectors. In order to find resistant material that replaces species or cultivars heavily affected by climatic change, breeding new, genetically diverse varieties of indigenous species, or testing of alien non-weedy species would be helpful.

With regard to their long lifecycles and to the tall shape of urban woody plants, the sustainable use of foreign material is mainly a question of accepting responsibility for the future impact of present-day activities. The increasing global impact of invasive tree weeds within cities, and that of problematic species spread by the urban citizens into ecosystems in the vicinity, are of major concern on all continents. City dwellers and amateur gardeners, as well as urban authorities and companies, should become aware of their responsibility to avoid loss of urban biodiversity and unnecessary introductions of invasive species. Recently, a lot of information resources have been developed for each kind of plant user, at different levels of professionalism.

References

Blackburn, T. M., Essl, F., Evans, T., *et al.* (2014). A Unified Classification of Alien Species Based on the Magnitude of their Environmental Impacts. *PLoS Biology* **12**, 36–49.

Carrillo-Gavilán, M.A. and Vilà, M. (2010). Little evidence of invasion by alien conifers in Europe. *Diversity and Distributions* **16**, 203–213.

Chace, T.D. (2013). *How to Eradicate Invasive Plants*. Timber Press, Portland, Oregon and London.

Colautti, R.I. and MacIsaac, H.J. (2004). A neutral terminology to define 'invasive' species. *Diversity and Distributions* **10**, 135–141.

DAISIE (ed., 2009). *Handbook of alien species in Europe/DAISIE*. Springer, Dordrecht/ Heidelberg.

de Oliveira, J.A.P., Balaban, O., Doll, C.N.H., Moreno-Penaranda, R., Gasparatos, A., Iossifova, D. and Suwa, A. (2011). Cities and biodiversity: Perspectives and governance challenges for implementing the convention on biological diversity (CBD) at the city level. *Biological Conservation* **144**, 1302–1313.

Elmqvist, T., Fragkias, M., Goodness, J., Güneralp, B., Marcotullio, P.J., McDonald, R.I., Parnell, S., Schewenius, M., Sendstad, M., Seto, K C. and Wilkinson, C. (eds, 2013). *Urbanization, Biodiversity and Ecosystem Services: Challenges and Opportunities/A Global Assessment*. Springer, Dordrecht.

Elton, C.S. (1958). *The Ecology of Invasions by Animals and Plants*. Methuen, London.

Genovesi, P. (2011). Eradication. In: Simberloff, D., Rejmánek, M. (eds.) *Encyclopedia of Biological Invasions*, pp. 198–203. University of California Press, Berkeley/ Los Angeles/ London.

Hejda, M., Pyšek, P. and Jarošík, V. (2009). Impact of invasive plants on the species richness, diversity and composition of invaded communities. *Journal of Ecology* **97**, 393–403.

Hope, D., Gries, C., Zhu, W. X., Fagan, W.F., Redman, C.L., Grimm, N.B., Nelson, A.L., Martin, C. and Kinzig, A. (2003). Socioeconomics drive urban plant diversity. *Proceedings of the National Academy of Sciences of the United States of America* **100**, 8788–8792.

Kleinschmit, J.R.G., Kownatzki, D. and Gregorius, H.R. (2004). Adaptational characteristics of autochthonous populations – consequences for provenance delineation. *Forest Ecology and Management* **197**, 213–224.

Kowarik, I. (2011). Novel urban ecosystems, biodiversity, and conservation. *Environmental Pollution* **159**, 1974–1983.

Kowarik, I. and Pyšek, P. (2012). The first steps towards unifying concepts in invasion ecology were made one hundred years ago: revisiting the work of the Swiss botanist Albert Thellung. *Diversity and Distributions* **18**, 1243–1252.

Kowarik, I. and Rabitsch, W. (2010). *Biologische Invasionen – Neophyten und Neozoen in Mitteleuropa*. Eugen Ulmer, Stuttgart.

Kowarik, I. and Säumel, I. (2007). Biological flora of Central Europe: *Ailanthus altissima* (Mill.) Swingle. *Perspectives in Plant Ecology Evolution and Systematics* **8**, 207–237.

Kühn, I., Brandenburg, M. and Klotz, S. (2004). Why do alien plant species that reproduce in natural habitats occur more frequently? *Diversity and Distributions* **10**, 417–425.

Kumschick, S., Gaertner, M., Vilà, M. *et al.* (2015). Ecological Impacts of Alien Species: Quantification, Scope, Caveats, and Recommendations. *Bioscience* **65**, 55–63.

McGregor, K.F., Watt, M.S., Hulme, P.E. and Duncan, R.P. (2012). What determines pine naturalization: species traits, climate suitability or forestry use? *Diversity and Distributions* **18**, 1013–1023.

Mooney, H. (2001). Invasive Alien Species – The Nature of the Problem. In: Secretariat of the Convention on Biological Diversity (SCBD) (ed.) *Assessment and management of alien species that threaten ecosystems, habitats and species*. Abstracts of keynote addresses and posters presented at the sixth meeting of the Subsidiary Body on Scientific, Technical and Technological Advice, held from 12 to 16 March 2001 in Montreal, Canada. SCBD, pp. 1–2.

Parker, J.D., Torchin, M.E., Hufbauer, R.A. *et al.* (2013). Do invasive species perform better in their new ranges? *Ecology* **94**, 985–994.

Petit, R.J., Bialozyt, R., Garnier-Gere, P. and Hampe, A. (2004). Ecology and genetics of tree invasions: from recent introductions to Quaternary migrations. *Forest Ecology and Management* **197**, 117–137.

Pyšek, P. (1998). Alien and native species in Central European urban floras: a quantitative comparison. *Journal of Biogeography* **25**, 155–163.

Pyšek, P., Richardson, D. M., Rejmánek, M., Webster, G.L., Williamson, M. and Kirschner, J. (2004). Alien plants in checklists and floras: towards better communication between taxonomists and ecologists. *Taxon* **53**, 131–143.

Pyšek, P., Jarošík, V., Pergl, J., Randall, R., Chytrý, M., Kühn, I., Tichý, L., Danihelka, J., Chrtek jun., J. and Sádlo, J. (2009a). The global invasion success of Central European plants is related to distribution characteristics in their native range and species traits. *Diversity and Distributions* **15**, 891–903.

Pyšek, P., Křivánek, M. and Jarošík, V. (2009b). Planting intensity, residence time, and species traits determine invasion success of alien woody species. *Ecology* **90**, 2734–2744.

Radtke, A., Ambrass, S., Zerbe, S., Tonon, G., Fontana, V. and Ammer, C. (2013). Traditional coppice forest management drives the invasion of *Ailanthus altissima* and *Robinia pseudoacacia* into deciduous forests. *Forest Ecology and Management* **291**, 308–317.

Rejmánek, M., Richardson, D.M. and Pyšek, P. (2012). Plant Invasions and Invasibility of Plant Communities. In: van der Maarel, E. and Franklin, J. (eds.) *Vegetation Ecology*, 2nd edition, pp. 387–424. Wiley-Blackwell.

Richardson, D.M. and Rejmanek, M. (2011). Trees and shrubs as invasive alien species – a global review. *Diversity and Distributions* **17**, 788–809.

Richardson, D.M., Pyšek, P., Rejmánek, M., Barbour, M.G., Panetta, F.D. and West, C.J. (2000). Naturalization and invasion of alien plants: concepts and definitions. *Diversity and Distributions* **6**, 93–107.

Roloff, A., Korn, S. and Gillner, S. (2009). The Climate-Species-Matrix to select tree species for urban habitats considering climate change. *Urban Forestry & Urban Greening* **8**, 295–308.

Ross, M.A. and Lembi, C.A. (2009). *Applied Weed Science: Including the Ecology and Management of Invasive Plants*. Pearson Prentice Hall, Upper Saddle River, New Jersey and Columbus, Ohio.

Schmidt, P.A. and Wilhelm, E.-G. (1995). Die einheimische Gehölzflora – Ein Überblick. Artbestand und Lebensformspektrum der einheimischen Gehölzflora. *Beiträge zur Gehölzkunde* **1995**, 50–75.

Simberloff, D., Alexander, J., Allendorf, F., Aronson, J., Antunes, P.M. *et al.* (2011). Non-natives: 141 scientists object. *Nature* **475**, 36–36.

Sukopp, H. (1998). On the Study of Anthropogenic Plant Migrations in Central Europe. In: Starfinger, U., Edwards, K., Kowarik, I. and Williamson, M. (eds.) *Plant invasions/ecological mechanisms and human responses*, pp. 43–56. Backhuys Publishers, Leiden, The Nehterlands.

van der Mijnsbrugge, K., Bischoff, A. and Smith, B. (2010). A question of origin: Where and how to collect seed for ecological restoration. *Basic and Applied Ecology* **11**, 300–311.

von der Lippe, M. and Kowarik, I. (2012). Interactions between propagule pressure and seed traits shape human-mediated seed dispersal along roads. *Perspectives in Plant Ecology Evolution and Systematics* **14**, 123–130.

CHAPTER 15

Criteria for species selection: Development of a database for urban trees

Sten Gillner[1], Mathias Hofmann[2], Andreas Tharang[1] and Juliane Vogt[1]

[1] *Technische Universität Dresden, Tharandt, Germany*
[2] *Snow and Landscape Research WSL, Social Sciences in Landscape Research Subunit, Birmensdorf, Switzerland*

15.1 Introduction

Trees and shrubs are key elements contributing to a healthy and attractive urban environment. Thus, the implementation of urban green is one effective way to counteract negative microclimatic conditions and air pollution. Tree canopies absorb and reflect solar radiation received by impervious urban materials, and lower air and surface temperatures by tree shading and evapotranspiration (Armson *et al.*, 2012; Douglas and James, 2015). Surface temperatures of an urban part of Dresden/Germany are shown in the thermography picture in Figure 15.1, demonstrating well the higher surface temperatures on concrete or asphalt covered surfaces (e.g., the bridge), compared to vegetated areas such as lawns or green spaces covered with trees along the riversides. They also reduce strong urban winds, control humidity, and can even contribute to flood prevention and erosion control (Tyrväinen *et al.*, 2005).

In many countries, programs have been implemented and decisions have been taken to increase urban tree population (Konijnendijk, 2003). Although management practices and tree establishment have improved (Dujesiefken *et al.*, 2005; Sieghardt *et al.*, 2005), the specific requirements and the knowledge about plant stress level is often left in the background. Municipalities, city planners and managers are frequently overwhelmed by the abundance of factors, and positive and negative effects, linked to the plantation of trees and shrubs (see Chapter 1). The selection of suitable tree species by databases facilitates the decision for the right species, adapted under the specific environmental conditions. Furthermore, choosing the optimal tree – one that is well-adapted to the site – enables fast tree establishment, gains the highest benefits and causes the lowest disruption to the inhabitants.

The generally used term "urban sites" obscures the broad range of different soil and microclimatic conditions that result from the history of building and settlement structure. The settlement history in cities caused destruction, demolition and reconstruction of buildings and infrastructures, and led to strongly heterogeneous environmental conditions. In many urban areas, space is limited (Roberts *et al.*, 2006). Consequently, the size of planting site and crown space are restricted for optimal growth. The specific site and microclimatic environment at paved sites, with higher

Urban Tree Management: For the Sustainable Development of Green Cities, First Edition. Edited by Andreas Roloff.
© 2016 John Wiley & Sons, Ltd. Published 2016 by John Wiley & Sons, Ltd.

Figure 15.1 Thermography indicating the surface temperatures on a hot summer day during August 2013. Vegetated areas show significantly lower surface temperatures, marked by different colors.

temperatures and lower water supply, result in stressful growth conditions (Gillner *et al.*, 2014). However, in contrast to this, at other sites – for example, urban parks and gardens near streams or ponds – water and nutrient supplies may be plentiful, offering better growth conditions. Harmful effects may also include exposure to air pollutants, de-icing salt and vandalism, which lower the disease resistance of urban trees (Sæbø *et al.*, 2003; Ascher, 2005).

Therefore, a broad range of factors, including climatic conditions, soil characteristics, space for roots and crowns, and possible residential preferences, must be considered for urban tree placement (Wu *et al.*, 2008). For planned tree sites, planning should consider suitable tree species that is selected for the given growth conditions, and which meets human demands. Tree selection and arrangement must be embedded in the composition of structural-spatial constructions types and historical quarter development, according to urban structure types and current environmental conditions.

When the urban area is simplified to one homogeneous stressful tree habitat, few tree species meet all possible requirements, and this results in low tree biodiversity – in most cities, for example, a small number of species dominate the total tree population (Sæbø *et al.*, 2003, 2005).

Only richly structured and composed urban green spaces that are based on ecological principles serve as habitats for associated fauna (FLL, 2008, 2010) and prevent widespread deaths by species-specific diseases. To counteract low tree diversity, it is important to differentiate urban areas into a mosaic of distinct tree habitat types that match human requirements for trees and tree growth conditions.

In recent years, databases for the selection of woody plants have been developed. However, databases that are focused on botanical attributes frequently omit detailed site requirements. Others ignore the possible negative aspects deriving from the plantation of trees and shrubs – for example, the spread of toxic plant material, the allergic potential or limb breakage. Consequently, a useful database for woody plants should include:

a) a high number of trees and shrubs to assure biodiversity;

b) detailed species descriptions;

c) information about site and environmental requirements of the species;

d) information about negative effects from species;

e) fast processing input mask with default presets for specific urban tree location types.

This chapter provides instructions and basic knowledge for criteria for species selection and describes the steps for the development of a database for urban trees. As an illustration, it uses the database "*Citree*" (https://141.30.134.137/citree/).

15.2 Species description, growth conditions, and risks related to species use

15.2.1 Data pool and nomenclature

The plantation of woody species requires very accurate information about growth development, plant characteristics and site requirements. Growth characteristics (e.g., crown shape and tree heights) differ strongly, depending on the provenance of a species (Arend *et al.*, 2011). In contrast to this, plantation of cultivars, hybrids and varieties offer several advantages – a better assessment of future growth development, such as planting trees with small heights in areas with overhead lines – as well as a better risk assessment, such as excluding negative effects to cars and humans by planting non-fruiting trees at parking lots and pedestrian zones. The database *Citree* contains species that are native and non-native to the temperate zones of Europe, but also cultivars, varieties, and hybrids.

The botanical nomenclature of Roloff and Bärtels (2014), which is under consideration by the International Code of Nomenclature for plants (ICN), was used to name species and subspecies, varieties, cultivars and hybrids. Assuring an adequately large data pool, all 243 species of the Climate Species Matrix (Roloff *et al.*, 2009) and, additionally, species, subspecies, varieties, cultivars and hybrids, were used for classification.

15.2.2 Literature review and evaluation

For the literature review for the *Citree* database, more than 75 national and international books, scientific journals and online plant databases were used to obtain detailed information for every aspect of each plant. Information was selected and systematically categorized. Where the data in several publications was based on one original source, then this was exclusively used.

The most important issue when it comes to obtaining reliable information is the evaluation of a variety of sources, including data from field studies, results from laboratory research and botanical literature. The information on the scales of measurement used in the original sources was elementary for qualitative evaluation. For the majority of ratio scaled data (e.g., mean tree height, mean crown width) or nominal scaled data (e.g., type allowing only yes or no answers), evaluation was conducted on the basis of frequency of occurrence.

Much more difficult was an evaluation of the ordinal scaled data, since tolerances and sensitivities (e.g., for waterlogging or soil compaction) were frequently described as "not tolerant", "very tolerant", "tolerant", "little tolerant" or "quite tolerant". In most of these cases, the data gave similar but never identical information, so that different descriptions of data were transformed into an own categorization for every source. Scientific and practical knowledge of plant specialists was used to obtain a better foundation for subsequent evaluation, and additional information for the species and attributes. An interim evaluation of all attributes was made in eight workshops, each lasting one day. In a final step, the interim evaluation for every category and every attribute was determined.

15.2.3 Structure of the database

Citree is structured into the following main categories: "plant characteristics", "natural distribution and requirements", and "hazard and disease". The main category tree appearance consists of six subcategories: "habitus", "leaf", "flower", "fruit", "prickles or thorns" and "surface-near roots" (see Table 15.1). In most subcategories, several aspects can be found. For example, mean and maximum tree height, crown width and shape and growth rate belong to the subcategory "habitus".

The second main category, "natural distribution and requirements" comprises two subcategories. The subcategory "natural distribution" provides a systematic and detailed description of the habitat, soil and climate factors of a plant, as described by Kiermeyer (1995) and Roloff and Bärtels (2014). The plant's sensitivities and tolerances, as well as soil and climatic requirements, are given in the subcategory "requirements" (Table 15.1). This category includes aspects that facilitate the assessment of species suitability under specific urban environments – for example, light requirement, sensitivity to soil compaction, pH-range of the soil, soil moisture range or de-icing salt tolerance. Special consideration was given to the usability of a plant at street tree sites, because of their dominant position in urban settlements and because they have to fulfill special requirements. They should have a mean to high tolerance to urban microclimatic conditions, as well as to smoke and industrial emissions. The minimum mean height should be over 5 m. Interference from fruit fall and coniferousness disqualify these plants as suitable urban street tree species.

The main category "hazard and disease" was subdivided in the three subcategories: "interferences", "danger", and "diseases and pests". Negative effects from woody plants – for example bad odors – are listed in the subcategory "interferences", while danger effects (for example, risk of toxicity and allergenic reactions) are included in the subcategory "danger". The presence of thorns, spines, and prickles and surface-near rooting may also have negative effects on humans and infrastructure. However, both these subcategories are listed in the main category "plant characteristics", since they are more strongly related to this main category.

15.3 Urban Tree Location Categorization

Urban structure types, which were determined according to real urban environments and are used for planning processes, were differentiated according to usage, function and construction purposes, especially for urban trees (Figure 15.2).

Some tree selection characteristics are independent of the urban structure type include natural soil conditions (e.g., pH), light availability, winter hardiness zone or late frost risk, whereas other characteristics are specific to particular urban locations. For example, light availability is strongly dependent on both neighboring trees and surrounding buildings that block sunlight.

Table 15.1 Structure of database.

Main category: Plant Characteristics						
Subcategories	Habitus	Leaf	Flower	Fruit	Prickles or thorns	Surface-near rooting
Aspects	• tree, shrub • trunk: single, multi-stemmed • growth: direction, rate • height: mean, max • crown: shape size, density	• evergreen vs. deciduous • needle vs. leaf • shape • size • autumn color	• color • fragrance • flowering period • inflorescence • honey plant trees	• color • shape • edibility • bird food trees		

Main category: Natural Distribution and Requirements		
Subcategories	Natural distribution	Requirements
Aspects	• area of origin • habitat, soil and climate description (Roloff and Bärtels, 2014) • hardiness: plant hardiness zone, late frost risk, winter hardiness (Roloff et al., 2009)	• light • soil: pH range, moisture, depth, compaction • waterlogging tolerance • drought tolerance (Roloff et al., 2009) • heat tolerance • de-icing salt tolerance • tolerance to smoke and industrial emission • management intensity • usability as street tree

Main category: Hazard and Disease		
Subcategories	Interferences	Danger	Diseases and pests
Aspects	• root damage • fruit fall • odor	• allergic potential • pollinat on period • presence of toxic plant parts • limb breakage • invasive potential	• pathogens (bacteria, viruses, fungi, animals)

Figure 15.2 *Liquidambar styraciflua* growing in two different urban tree locations along streets (left) and within a park (right).

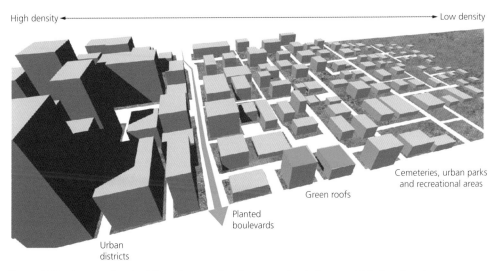

Figure 15.3 Diagram of scale shift and constraints for vegetation in low- to high-density urban areas (Illustration by courtesy of Keith Van de Riet).

Besides site conditions which vary according to urban structure types, some characteristics can be categorized according to usage types. In the following, we combine these characteristic categories and classify them as urban tree location types, which are exemplarily illustrated in Figure 15.3.

15.3.1 Urban tree location types

Road traffic areas consist of different street types (e.g., main, side or residential streets), public places, parking spaces and intersections. Street trees structure and connect urban settlements, provide shelter from direct sunlight, cool the air, and filter dust and noise from traffic. Plants which may be considered as street trees have to fulfill a wide range of special requirements. Street tree locations are generally expected to be relatively plant-hostile (Meyer, 1982), due to their low atmospheric humidity, high air pollutant concentrations and noticeable warming (Sukopp and Weiler, 1988). Moreover, tree pit space is often highly limited and characterized by soil compaction, caused by mechanical loads, and sealing or contamination that is caused by fuel or de-icing salt (Meyer, 1982; Roberts *et al.*, 2006; Sukopp and Weiler, 1988). Trees along traffic areas must have a specific height to meet utility clearance requirements (FLL, 2010; Meyer, 1982).

Therefore, the mitigation requirements to be considered in the selection of suitable species, subspecies, varieties, cultivars and hybrids in these locations are very diverse. Most can be met by thermophilic single-stem plants, which have a high level of stability, little vulnerability against wind breakage, a certain tolerance to pruning and no interfering fruit fall (Gälzer, 2001). Furthermore, the plants should have a high resilience against pollution, de-icing salt and soil compaction, and tolerance of pH values up to 8. Aesthetic design criteria can be secondary considerations after required growth conditions.

Densely built-up areas include residential areas with specific construction types, such as block or linear development, and very diversely constructed inner cities (Sæbø *et al.*, 2003; Sukopp and Weiler, 1988). The characteristics of these tree locations largely overlap with the aforementioned traffic areas, although with less emphasis on complying with utility clearances and tolerance to de-icing salt. Specifically, the stress factors for trees, as well as soil and microclimate conditions in these areas, are very similar to traffic areas – or even higher. More focus must be given on interferences, such as building darkening, allergic and toxic potentials, and unpleasant flower and fruit smells. Additionally, the well-being and aesthetic effects of trees have a greater emphasis within densely built-up residential areas.

Green roofs and container plants are often used to create extra space without using additional land (FLL, 2008), which is convenient in densely built-up areas with no available free space. Trees located on rooftops, on top of underground parking areas, or in containers, have limited rooting space; therefore, small-crowned trees should be used in these locations, because large trees are not suitable for the growth conditions (Meyer, 1982). The restricted soil depth holds a high risk of soil drought. Therefore, drought-tolerant species should be planted in these areas. Maintenance is fairly extensive – using porous and water-retaining soil substrate, trees rely on regular fertilization and watering (Meyer, 1982).

Sparsely built-up urban areas can be subdivided into sparsely populated residential areas (e.g., detached and semi-detached houses, front gardens), recreation grounds, and green corridors without motorized traffic (e.g., Sukopp and Weiler, 1988). On the one hand, these areas are characterized by lower stress caused by reduced anthropogenic-driven influences compared with traffic areas or densely built-up areas. However, on the other hand, soil compaction by mechanical loads and pollutants might be important in sparsely built-up areas. These negative effects can be partly balanced by extensive maintenance, such as artificial fertilization, irrigation or thinning activities by competitors (Meyer, 1982). Nevertheless, these tree locations must fulfill certain functions, compared with street trees, such as avoiding risks to the well-being of people via toxicity, allergen potential or breaking branches (Gälzer, 2001; Sæbø *et al.*, 2003).

Gälzer (2001) listed several varieties of **industrial and commercial green areas**, including urban green spaces around offices, administration buildings, trading enterprises, manufacturing factories, waste disposal companies, and within technology centers or science parks. In general, these locations are often characterized by warming, soil compaction, and site-specific pollution (Sæbø *et al.*, 2003), due to the relatively high sealing rate. Moreover, industrial areas exhibit particularly high emissions. Tree planting should be integrated besides construction facilities in the surrounding area, and provide recreation and protection from emissions for workers and visitors (Gälzer, 2001). In summary, trees on industrial and commercial sites should be predominantly robust and tolerant to soil compaction, pollution and heat, and aesthetic aspects affect the well-being of the users.

Public parks, gardens and cemeteries often follow a preconceived design concept. Recreational activities as well as the experience of nature and landscape represent important targets of park planning. Borchardt (2010) listed the diverse aspects of the use requirements for different urban greens; areas with predominantly ornamental characteristics, such as parks, gardens and other green spaces, have high maintenance costs. The aesthetic quality is partially created by a high number of non-native species. Cemeteries and memorial places are commonly characterized by a favorable microclimate, with humus accumulation and artificial irrigation systems, which favor moist-meadow and riverbank species (Sukopp and Weiler, 1988). These locations are richly structured, with large trees, shrubs and hedges. Deep-rooting plants should be preferred to plants with a shallow root system, due to grave excavation in these areas (Gälzer, 2001).

Trees along **waterways, ports and channels** often play an important role for stabilizing banks with their root system. These areas are characterized by reduced climate extremes, humidity, and soils that might be eutrophic and polluted (Sukopp and Weiler, 1988) due to shipping activities or waste water disposal. Additionally, urban rivers and channels might serve as dispersal corridors for invasive tree species (Säumel and Kowarik, 2010). Therefore, eutrophic, waterlogging-tolerant native or non-invasive plants with good anchorage should be primarily considered for green spaces along waterways.

Urban forests can consist of remnants of natural forests or new planting after land use changes. Generally, urban forests evince a relatively low stress level; however, these areas might be affected by soil compaction, pollutants (depending on the specific land usage) and the history of the location (Meyer, 1982; Sæbø *et al.*, 2003). The management of urban forests is similar to non-urban forests – that is, successional processes, pest control and thinning that affects the timber quality and growth rate are all important (Meyer, 1982; Sæbø *et al.*, 2003). Urban forests are among the few urban tree locations that might be interesting for timber production. Nevertheless, recreational and sporting activities within woodlands with extended foot and cycle paths outweigh the potential utilization of wood for timber production (Gälzer, 2001).

Tree locations in **renaturation areas and on derelict land** emerge either by natural succession processes resulting from abandonment (rural areas or derelict land) or muss hier nicht ein "by" oder "from" dazwischen (je nachdem, auf welchen der vorangehenden Teile des Satzes sich das bezieht)? having been specifically laid out for renaturation purposes after being abandoned (i.e., land use change). In general, native, persistent, low-maintenance plants, with rich vertical and horizontal structures, should be used to create species-rich habitats (FLL, 1999; Matsuoka and Kaplan, 2008). These ruderal sites have a relatively benign microclimate, although they might be contaminated by former usage (e.g., being former industrial areas or disposal areas), and the soil may contain a high amount of rocks or metals (Sukopp and Weiler, 1988). Low, competitive fast-growing pioneer species are abundant on these sites, due to the relatively young developmental stage of these sites (Sukopp and Wittig, 1998; Gälzer, 2001).

Some derelict areas are subjected to interim uses in order to lower maintenance costs while bridging the time until the land is reused as commercial, industrial or residential areas. Other areas are subject to planning to create human recreation areas (e.g., the countryside park Duisburg-Nord/Germany (Hertlein-Rieder, 2007)).

Allotment gardens are predominantly designed to grow fruits and vegetables, and for private recreation purposes (see Chapter 18). Here, the tree selection process focuses on fruit trees for cultivating food. These locations are commonly characterized by good maintenance, with additional irrigation and fertilization. However, problems with eutrophication may occur.

15.3.2 Specific demographic groups

For people with a relatively small reach of action, areas close to housing or working places are extremely important for recreation. In this section, therefore, we focus on the special needs of specific demographic groups, including children and elderly people. These people are likely to be more vulnerable to the negative environmental impacts of specific urban climate and emissions, and this poses a challenge to create urban green spaces that meet the special needs of sensitive demographic groups, particularly within the residential environment.

Trees provide nature experience for urban children, and are used for playing – for example, climbing, testing physical skills or creating tree houses (Smardon, 1988). Thereby, children improve their physical, mental and emotional health and well-being, as for example reducing asthma rates due to improved air quality by trees (Armour et al., 2012). Edible fruits from trees and naturally composed arrangement have been suggested to promote environmental education, and to provide nature experiences (Fellenberg, 1991; Matsuoka and Kaplan, 2008). Tree planting, and the associated wildlife, act as illustrative material, crafts, and as an experimentation and observation area for ecological processes (FLL, 1999). Shady trees support the well-being of people during hot and sunny days (Fellenberg, 1991). Toxic, aculeate and potentially allergenic trees, and trees with a high risk of breaking branches, should be avoided.

For the elderly aesthetic aspects become increasingly important (Fellenberg, 1991; Matsuoka and Kaplan, 2008). Contact with nature supports efforts to remain healthy and active (Armour et al., 2012). Urban tree plantings to the elderly should be designed with diverse leaf coloring, flowering and fruiting seasons (Fellenberg, 1991).

Based on the above description of the urban tree location types and characteristic requirements of trees, we developed a decision matrix which is integrated as a simplified fast processing input mask within the database (Figure 15.4).

Figure 15.4 shows that the requirements are quite high and diverse for some urban tree locations (e.g., traffic areas and densely built-up areas). Thus, many aspects must be considered when planting or maintaining trees in these locations. On the other hand, no specific priorities need to be set for urban forests. The developed decision matrix with the tree location classification simplifies and structures the selection of relevant aspects of urban species, and helps the users to obtain a fast overview (see Chapter 2).

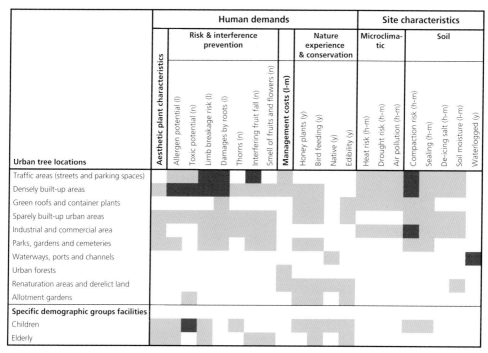

Figure 15.4 Decision matrix that combines the classified urban tree locations with the different aspects of human demands and required site conditions for urban trees. The boxes present the different levels of priority that should be considered when selecting trees in specific locations: dark green – fixed linkages with the highest priorities; light green – highlighted priorities that are not necessarily always related; white – no special priorities. Abbreviations are as follows: **y** – yes, **n** – no, **l** – low, **m** – medium, and **h** –high.

15.4 Psychological aspects of the database

15.4.1 User-based urban green space categorization

To estimate the relevance of the literature-based classification of tree locations for the everyday life of city residents, as described above, we devised a user-based classification of urban open spaces; in other words, a classification of urban spaces where trees or shrubs may grow. Using standard literature on urban green space planning (Gälzer, 2001), we collected labels that characterize specific urban open spaces, and added short descriptions. In an online study, volunteering city residents were asked to sort these labels according to subjectively perceived similarity, putting potentially similar spaces into the same group, while putting dissimilar spaces into different groups.

Using cluster analysis, we identified a number of groups of urban open spaces which are distinct from another in the view of their potential users. These groups were:
1 Private spaces (under the individual's personal control).
2 Spaces belonging to certain buildings (e.g., to children's facilities or to retirement homes).
3 Recreational spaces without access restrictions.
4 Recreational spaces with restricted access.

5 Vast expanses of land serving certain non-recreational purposes (e.g., landfills).
6 Smaller, functional spaces related to road or pedestrian traffic.
A number of parallels to the literature-based classification of tree locations are apparent, but there are also some differences. Both classifications feature a category of spaces related to traffic. Group 5 of the user-based classification has similarities to the category "renaturation and derelict land", but it also contains urban forests, which constitute a single category in the literature-based classification. Finally, the open-access "recreational" group 3 comprises spaces such as those in the category "parks, gardens, and cemeteries".

Another fundamental difference is that, in the user-based classification, the most important distinction is made between private and public spaces. This is not only visible in the clustering solution, but also becomes apparent when analyzing the data using multidimensional scaling (MDS) procedures. The latter yielded four important dimensions by which urban open spaces are distinguished (in order of importance):
1 Degree of privacy.
2 Vastness of the space.
3 Amount of traffic present at the space.
4 Availability of vegetation.
The importance of the degree of privacy may be an expression of the fundamental psychological human need for orientation and control (Grawe, 2007). The more private an open space can be considered, the more familiar one will be with it, the easier it is to orient oneself in it, and the more it will allow for personal control. Although private spaces are not usually subject to urban planning, the importance of privacy may be considered in designing public spaces; when designed accordingly, these spaces may also offer a certain degree of (albeit temporary) privacy.

15.4.2 Tree perception and tree preferences

An important psychological aspect in tree selection concerns possible preferences of city residents for certain trees in their living environments (e.g., for trees lining residential streets). No previous studies could be found which had systematically researched this question. In trying to answer it, we took a two-step approach.

First we identified tree features relevant in human perception regarding distinction among trees. To that end, lay participants completed a task in which they sorted computer-rendered images of trees according to perceived similarity. They were free to choose a similarity criterion meaningful to them. Using cluster analysis, we identified distinct tree categories:
1 Conifers.
2 Trees with a globular-shaped crown and a long trunk.
3 Trees with a globular-shaped crown and a short trunk.
4 Trees with an oval-shaped and an opaque crown.
5 Trees with an oval-shaped and a translucent crown.
To help interpretation of the underlying distinctions, we used multidimensional scaling techniques on the sorting data. We identified a number of relevant perceptual criteria:
1 The distinction between conifers and deciduous trees.
2 Crown shape (ratio of crown height to crown width).
3 Trunk length.
4 Ratio of the two-dimensional crown size to trunk length.
5 Crown opacity.

Figure 15.5 Tree preferences: Example for tree with low (*Pinus nigra*, left) and high preference (*Fraxinus excelsior*, right). (Illustration by courtesy of Tina Gerstenberg.)

In the second step, we researched the connection between these criteria and tree preferences. In a different study, we asked city residents to rate a set of tree photographs according to personal preference in their living environment. For these trees, typical values for the previously identified criteria were known. These criteria were used as possible predictors in linear regression analysis. Using a backwards variable elimination procedure, a model was identified which predicts tree preference using:

1 The distinction between conifers and deciduous trees.
2 Maximum tree height (trunk length + crown height).
3 Crown shape.

This means that large deciduous trees with wide crowns are most preferred. Figure 15.5 illustrates the two extremes of the preference spectrum: on the left, a comparatively small coniferous tree with a rather slender crown (*Pinus nigra*) was among the least preferred, while the large deciduous tree with its wider crown (*Fraxinus excelsior*) on the right was found to be among the highly preferred trees. We used the statistical preference prediction model to estimate preference values for all trees contained in the *Citree* database.

15.5 Application possibilities and limitation of use

For decision-makers such as landscape planners, the developed urbantree database with the different compounds (botanical, urban tree locations, ecosystem services, and psychological aspects) provides guidelines for the selection of trees, following the principle of having the right tree for the right place. This database is most suitable for future tree planting. One possible use for the database concerns the selection of genetically different tree species which share a similar appearance in regard to the tree characteristics most relevant to human perception. For example, this may be useful to achieve a harmonious overall appearance of the trees lining a certain street, while lowering the risk of all trees being harmed at the same time by pests or by changes in climatic conditions. However, the database may also apply to management purposes for existing trees. Consequently, tree removal that is triggered by disservices on human well-being and health, or on non-optimal tree establishment, growth and resilience conditions, can be avoided.

The database offers a wide range of possible applications for different users. However, future research should focus on several aspects only marginally covered in the database – for example, botanical features like the specific discoloration of trees and shrubs, or the lifespan of the plants. Requirements, tolerances and recommendations given in the database refer to northern and mid-latitudes. In more southern regions, this may lead to false conclusions. The degree of endangerment of species is not included, since it varies strongly, depending on the region.

Although the urban tree location types emphasize the risk and interference potential triggered by trees, some conflict of use cannot be completely eliminated, despite careful planning. Therefore, it is important to minimize these conflicts by skillfully selecting tree species. The FLL (2010) distinguishes between acceptable (shading or barrier effects, overhanging branches, falling leaves, flowers and fruits) and unacceptable (damage to foundations, pavement or pipelines by roots, damage to roofs or facades, obscuration, and hazards due to limb breakage) interferences. According to this distinction, priorities must be set for tree planting.

Nevertheless, Kirkpatrick *et al.* (2013) mentioned that even healthy trees are in jeopardy of removal when aesthetic and lifestyle preferences change. For urban tree location types, it is almost impossible to comply with subjective alterations in fashion to meet differing aesthetic demands. Therefore, decisions on tree selections will always be an individual process which require tailoring towards the specific circumstances. Nonethelss, we believe that databases such as *Citree* may be of help in this process.

15.6 Conclusions

Choosing suitable trees and shrubs can reduce microclimatic loads, and can have positive effects on human health as well as on residential satisfaction. Urban environments are highly variable regarding their soil and climatic conditions and their user requirements. Therefore, the plant selection has to incorporate location-specific factors. Climate change makes it even more important to select suitable species under changing growth conditions, to guarantee the successful establishment of future urban vegetation.

The specific characteristics of more than 390 woody plants were investigated, focusing on urban growth conditions. The large data pool, containing species, subspecies, varieties, hybrids and cultivars, assures a wide range of suitable plants for different demands. A database was developed containing several main categories: site characteristics and natural distribution; appearance; ecosystem services management activities; risks and interferences of urban woody plants; and human preferences.

Although local particularities such as management and policy regulations must always be observed, the database may be a valuable decision support system for urban tree management.

References

Arend, M., Kuster T., Gunthardt-Goerg M.S., *et al.* (2011). Provenance-specific growth responses to drought and air warming in three European oak species (Quercus robur, Q. petraea and Q. pubescens). *Tree Physiology* **31**, 287–297.

Armour, T., Job, M. and Canavan, R. (2012). *The benefits of large species trees in urban landscapes: A costing, design and management guide.* (No. C712). CIRIA.

Armson, D., Stringer, P. and Ennos, A.R. (2012). The effect of tree shade and grass on surface and globe temperatures in an urban area. *Urban Forestry and Urban Greening* **11**, 245–255.

Ascher, K. (2005). *The works: Anatomy of a city.* Penguin Press, Singapore.

Borchardt, W. (2010). *Handbuch Pflanzenverwendung im Garten-und Landschaftsbau.* Patzer Verlag, Berlin/Hannover.

Douglas, I. and James, P. (2015). *Urban Ecology, an Introduction.* Routledge, New York.

Dujesiefken, D., Drenou, Ch., Oven, P., *et al.* (2005). Arboricultural practices. In: Konijnendijk, C.C., Nilsson, K., Randrup, T.B., *et al.* (eds). *Urban forests and trees,* pp. 419–441. Springer, Berlin.

Fellenberg, G. (1991). *Lebensraum Stadt.* Verlag der Fachvereine, Zürich.

FLL (Forschungsgesellschaft Landschaftsentwicklung Landschaftsbau e.V.) (eds, 1999). *Leitfaden für die Planung, Ausführung und Pflege von funktionsgerechten Gehölzpflanzungen im besiedelten Bereich,* 2nd edition. FLL, Bonn.

FLL (Forschungsgesellschaft Landschaftsentwicklung Landschaftsbau e.V.) (eds, 2008). *Green roofing guideline,* 7th edition. FLL, Bonn.

FLL (Forschungsgesellschaft Landschaftsentwicklung Landschaftsbau e.V.) (eds, 2010). *Baumkontrollrichtlinien: Richtlinien für Regelkontrollen zur Überprüfung der Verkehrssicherheit von Bäumen,* 2nd edition. FLL, Bonn.

Gälzer, R. (2001). *Grünplanung für Städte Planung, Entwurf, Bau und Erhaltung.* Ulmer, Stuttgart.

Gillner, S., Bräuning, A. and Roloff, A. (2014). Dendrochronological analysis of urban trees: climate response and impact of drought on frequently used trees. *Trees* **28**, 1079–1093.

Hertlein-Rieder, V. (2007). Rheinpark Duisburg. In: Bund Deutscher Landschaftsarchitekten bdla (eds). *Insight out,* pp. 79–82. Birkhäuser, Berlin.

Kiermeyer, P. (1995). *Die Lebensbereiche der Gehölze.* BdB-Verlagsgesellschaft, Pinneberg.

Kirkpatrick, J.B., Davison, A. and Daniels, G.D. (2013). Sinners, scapegoats or fashion victims? Understanding the deaths of trees in the green city. *Geoforum* **48**, 165–176.

Konijnendijk, C.C. (2003). A decade of urban forestry in Europe. *Forest Policy and Economics* **5**, 173–186.

Matsuoka, R.H. and Kaplan, R. (2008). People needs in the urban landscape: analysis of landscape and urban planning contributions. *Landscape and Urban Planning* **84**, 7–19.

Meyer, F.H. (1982). *Bäume in der Stadt.* Ulmer, Stuttgart.

Roberts, J., Jackson, N. and Smith, M. (2006). *Tree Roots in the Built Environment.* The Stationery Office, Norwich.

Roloff, A. and Bärtels, A. (2014). *Flora der Gehölze,* 4th edition. Ulmer, Stuttgart.

Roloff, A., Korn, S. and Gillner, S. (2009). The Climate-Species-Matrix to select tree species for urban habitats considering climate change. *Urban Forestry & Urban Greening* **8**, 295–308.

Sæbø, A., Benedikz, T. and Randrup, T.B. (2003). Selection of trees for urban forestry in the Nordic countries. *Urban Forestry & Urban Greening* **2**, 101–114.

Sæbø, A., Zelimir, B., Ducatillion, C., *et al.* (2005). The selection of plant materials for street trees, park trees and urban woodlands. In: Konijnendijk C.C., Nilsson, K., Randrup, T.B., *et al.* (eds.) *Urban forests and trees,* pp. 257–280. Springer, Berlin.

Säumel, I., Kowarik, I. (2010). Urban rivers as dispersal corridors for primarily wind-dispersed invasive tree species. *Landscape and Urban Planning* **94**, 244–249.

Sieghardt, M., Mursch-Radlgruber, E., Paoletti, E., *et al.* (2005). The abiotic urban environment: impact of urban growing conditions on urban vegetation. In: Konijnendijk, C.C., Nilsson, K., Randrup, T.B., *et al.* (eds.) *Urban forests and trees*, pp. 281–323. Springer, Berlin.

Smardon, R. (1988). Perception and aesthetics of the urban-environment – review of the role of vegetation. *Landscape and Urban Planning* **15**, 85–106.

Sukopp, H. and Weiler, S. (1988). Biotope mapping and nature conservation strategies in urban areas of the federal republic of Germany. *Landscape and Urban Planning* **15**, 39–58.

Sukopp, H. and Wittig, R. (1998). *Stadtökologie ein Fachbuch für Studium und Praxis*, 2nd edition. Gustav Fischer, Stuttgart.

Tyrväinen, L., Pauleit, S., Seeland, K., *et al.* (2005). Benefits and uses of urban forests and tress. In: Konijnendijk C.C., Nilsson, K., Randrup, T.B., *et al.* (eds). *Urban forests and trees*, pp. 81–114. Springer, Berlin.

Wu, C., Xiao, Q. and McPherson, E.G. (2008). A method for locating potential tree-planting sites in urban areas: A case study of Los Angeles, USA. *Urban Forestry & Urban Greening* **7**, 65–76.

CHAPTER 16

Genetic aspects

Doris Krabel

Technische Universität Dresden, Tharandt, Germany

16.1 The problem of trees from a genetic point of view

Trees in urban areas fulfill manifold and ambitious functions. Some of these functions are ecological, but they are mainly aesthetical, cultural and social (see Chapter 1). No other urban element for design is operational in such a flexible way, and has an impact which increases with age, like the planting of trees. From the moment of installation a tree into an urban habitat, however, the tree is exposed to numerous rapidly changing environmental conditions and, in most cases, ones that have a negative effect. Plants living in a suboptimal environment suffer stress, with consequences of reduced vitality, increased susceptibility to diseases and a shortened lifespan.

Of course, in nature also, there exist numerous factors that cause stress to trees (see Chapter 5). However, during several tens of thousands of years of evolution, trees have developed mechanisms of adaptation which enable individual trees, as well as stands or populations, to survive these stressors (Figure 16.1).

Generally, adaptation to the environment can arise in two ways. Concerning its life processes, each individual has a wide physiological range of ways to react to environmental changes. For instance, as a reaction to decreasing groundwater level, plants can develop a more deep or denser root system (see Chapter 4). Alternatively, or in addition, the water demand can be decreased by physiological control loops.

Usually, different individuals within one species and the same population or stand react to different degrees of intensity. Single individuals might come cope better with drought than others, which better tolerate cold winter temperatures, while yet others are more sensitive to early or/and late frost events, increased salt concentration of the soil, or pollution of the atmosphere. However, there does not exist a "super individual" which tolerates all stress factors equally well. At this point, genetic aspects come into play. The frame in which a plant can react individually to environmental changes is mostly determined by its genetic constitution (reaction norm) and, thus, by the dissemination of hereditary factors from its ancestors. Hence, a group or a population is able to react to stress factors far beyond the possibilities of a single individual. The group needs a certain genetic variability, which we call genetic adaptedness.

Considering that environmental conditions are naturally never constant over time, but change, especially under human impact, trees need "down locks" for future environmental conditions which may be actually unknown or irrelevant. In this case, we mean (genetic) adaptability. "Adaptedness" implies a high likelihood of survival in a given

Urban Tree Management: For the Sustainable Development of Green Cities, First Edition. Edited by Andreas Roloff.
© 2016 John Wiley & Sons, Ltd. Published 2016 by John Wiley & Sons, Ltd.

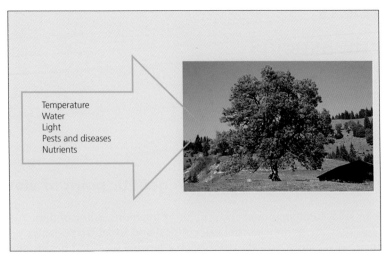

Temperature
Water
Light
Pests and diseases
Nutrients

Figure 16.1 Environmental (stress) conditions. Environmental factors stressing trees under natural conditions. (Photograph by S. Herzog.)

environment, whereas "adaptability" is the ability to react to changing environmental conditions.

Due to their long life cycles and lack of mobility, trees are, more than other groups of organisms, dependent on genetic adaptability. On the level of an individual plant, adaptability is expressed by adjusting the phenotype as a direct response to the changed environment (or parts of the environment). This ability is also called phenotypic plasticity (Schlichting, 2002), and it is particularly pronounced in woody plants (Stettler and Bradshaw, 1994). These physiological, anatomical or morphological adaptations are probably manifested in mechanisms such as rolling in their leaves under drought or frost events (e.g., Rhododendron spp.), the development of reaction wood reflecting mechanical strain, adaptation of the metabolism to cold temperatures, or the activation of specific genes which induce the production of substances for defense against pathogens (e.g., Parson *et al.*, 1989).

On the population level, adaptability and adaptedness to local conditions is manifested in genetic structures, which means the occurrence (qualitative and quantitative) of specific variants (genotypes) and their distribution in a population.

As a consequence, a population may differentiate into subpopulations with various ecotypes (as a result of selection processes) that are morphologically and/or physiologically adapted to sometimes highly specific ecological conditions (Krabel, 2011). Generally, the process of adaptation takes numerous generations, and is based on a permanent adjustment between environment and genetic variability within a population (Figure 16.2).

The process of adaptation for natural habitats could be also suitable for trees in urban environments. However, the problem is that urban trees are placed in artificial habitats, with environmental factors which are extremely and intensively influenced by humans. It can be assumed that, in urban areas, few tree populations exist which have completed a long-lasting adaption process to the site they are actually living in. Another problem in this context is that plant material chosen for urban greening usually originates from nurseries. Here, the plants were cultivated without natural selection, but selected on the basis of human (i.e., nursery) requirements (Figure 16.3).

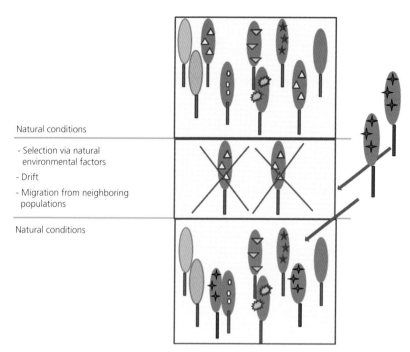

Figure 16.2 "Strategy" of adaptation on population level.

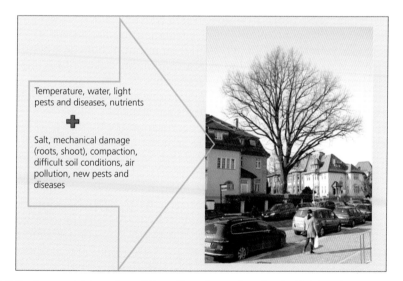

Temperature, water, light pests and diseases, nutrients

Salt, mechanical damage (roots, shoot), compaction, difficult soil conditions, air pollution, new pests and diseases

Figure 16.3 Environmental (stress) conditions: Stress factors in urban/artificial environment.

Those man-made selection criteria include, for instance, homogeneity of the plant material, vigor, shape of the crown, height of the plants, color of leaves and flowers, and so on. From the genetic point of view, only a very limited part of the total genetic variability of a (natural) population becomes represented in such a way. From those plants originating from a strongly limited gene pool, we now expect that they cope with every type of artificial urban environment.

16.2 Diversity, monoculture, variety and clones – some general comments

A generally used definition of the term "biodiversity", which was published by the United States Congress Office of Technology Assessment in 1987, takes account of the different levels which are encompassed by this term (Gaston, 1996). Thus, diversity can be regarded on the level of ecosystems, or species, or populations. The latter is especially essential for the long-term survival of a species.

In ecology, the term "diversity" is mostly used in a more general sense, and as a synonym for variation. In population genetics, diversity characterizes the genetic heterogeneity of populations or sub-populations, but based on the relative frequency distribution of genetic variants on a specific gene locus.

The term "urban tree population", as it is used in the following chapters, characterizes a set of trees of different species. Thus, in the genetic sense, the term is not correctly applied. Correctly, a population includes individuals of one community of reproduction, which means that each pair of individuals has at least the theoretical opportunity to produce one common descendent.

The term "monoculture" describes the cultivation of a large number (hundreds, thousands, millions) of plants of one single species, covering a large area. Monoculture does not necessarily mean genetic uniformity, because plants within a monoculture can be genetically different (different genotypes). Examples for large monocultures can be found in agriculture, where huge areas are covered with one crop species. In this context, especially in connection with trees, the term "plantation" is often used.

Plantations are typically (but not necessarily) monocultures of one and the same tree species, which have been established to gain a continuous yield, with economic objectives. These plantations often consist of only few clones. A clone is genetically completely homogeneous, which means that plants from one clone are absolutely uniform, from the apical meristem to the root tips. Generally, trees are cloned by taking cuttings (parts of the plant, such as shoot, root, meristem or embryo) from the mother tree, regenerating the roots and shoots, and cultivating them as new daughter plants (ramets). Plantings with a large number of individuals of a few clones, or even only one, on a huge area represent the extreme case of a monoculture.

"Cultivars" or "varieties" represent groups of plants which can be characterized by a defined trait (e.g., color of leaves, shape of flower), and they differ in such a manner from other groups of the same species (Roloff and Bärtels, 2014). This trait characterizes the cultivar even after propagation (e.g., grafting onto a stock or rooting of plant parts), and all the daughter plants show the same phenotype. Cultivars are not necessarily clones, but if a cultivar is propagated vegetatively from one mother plant, then the cultivar represents a clone. In the case of grafted cultivars, the graft should be rooted to gain a clone. However, all it takes to make a clone a cultivar is the application of a new name (Santamour, 1990).

Grafted plants belonging to the same variety are generally genetically uniform (same genotype) in the aboveground part, because they originate from one single mother individual. This means that, theoretically, these plants have the same phenotypic plasticity with regard to biotic and abiotic environmental conditions. To what extent the stock (roots and small part of the shoot), which is genetically diverse when it is originating from seeds (different genotypes), influences the individual phenotypic plasticity of the above-ground part of the plant is not yet fully understood. However, it can be assumed that the impact of substances produced/transported by/via the rootstock is not particularly drastic, because the distinctive characteristics of the cultivar are still preserved (Santamour 1990).

The advantage of using varieties is that these plants show reliably specific traits. Thus, it can be predicted how these plants perform, with respect to color, shape, form and size of the leaves, and so on. Advantages and disadvantages of the specific varieties are known to guarantee a specific quality, and this is even more true when a "proven variety" is used. Another advantage of grafted plants is that they are generally good compartmental-izing plants. Santamour (1990) justifies this statement by the fact that when screening plant material for grafting unintentionally, those plants are selected which tend easily to regenerate wound callus (good compartmentalizing trees) and, in this respect, the plants are especially suited for grafting.

A disadvantage arises when a late incompatibility reaction appears and the tissues of stock and graft do not really harmonize with each other. This phenomenon more likely occurs for new varieties, with which the nurseries do not have experience on older trees.

When choosing a large number of clonal plants, the criterion of homogeneity as an aesthetic requirement usually plays an essential role. The advantage is, as it was already mentioned for varieties, that the material is more or less well known.

16.3 The risk of missing diversity

Genetically, homogenous collections of large numbers of plants on a huge area are generally risky, because these cultures are genetically not very variable. To express it using the terms "adaptability" and "adaptedness": the plants may be sufficiently adapted to present environmental conditions (otherwise they will not grow and survive), but the adaptability to changing environmental conditions is quite limited. This means that the reaction potential of these plants as a whole is rather limited regarding biotic (e.g., insects, fungi) and abiotic risks (e.g., drought, salt concentration, compaction, root dam-age, temperature extremes) (Figure 16.4).

The following example may illustrate the problem: an avenue of a larger number of trees is planned for the next one hundred years, and one variety is selected due to its aesthetic function and tolerance to road salt. Now, if it is assumed that, under urban site conditions, the environmental situation changes relatively quickly (e.g., air pollution, increased dust formation, reduced water supply by sealing), with extreme values from time to time, then one runs an incalculable risk if clonal plant material has been used. It might be that the total plant population, or at least a large part of the avenue's trees, is not adapted to the conditions likely to prevail in 50, 60 or 100 years. As a result, early failure of not only some trees, but the total establishment, might be rather possible (Krabel and Herzog, 2008; Figure 16.5).

At this point the question arises: how could plantings of old and impressive trees survive up to now? Studies on the genetic constitution of residues of 250 year old lime tree avenues in two German cities showed that homogenous-appearing plantings are not necessarily genetically homogenous. For example, one avenue consisted of 199 trees, representing 104 different genotypes. The rest of the plants belonged to a few different clones (same genotype). In the latter case, it could be conjectured that these

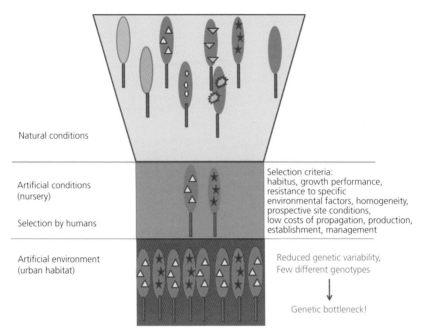

Figure 16.4 Scheme presenting the genetic bottleneck (red background) which is created by the selection of only few genotypes.

Figure 16.5 Unfavorable environmental site conditions are a poor basis for urban trees having a long life. (Photograph by S. Herzog)

clonal plants originated from different delivery batches, which were used for replenishment. Additionally, the study showed that a small number of trees were grafted, because root and stem were not genetically identical. For the 104 different genotypes, it can be assumed that, 250 years ago, mainly local seedlings from the forests, with comparably high genetic variability, were used for the establishment. The homogenous appearance of the avenue has been generated by cutting measures (Krabel and Herzog, 2008).

In general, not only in urban areas, it can be assumed that plantings or populations which lack diversity are more susceptible to calamities than populations with high diversity. Due to their long generation cycle, compared to their parasites and pathogens, trees are not able to react to changed situations by the formation of protective mechanisms. While herbaceous plants produce a new generation with probably resistant or less sensitive offspring in the following generation, the parasite generations inevitably encounter the same tree generation or same individuals year after year and can, thus, easily adapt to the host.

On the other hand, the survival of the tree population depends on the number of resistant genotypes available in the present population. An adaptation against present pathogens is only possible by future populations within decades. A targeted selection of trees resistant against specific pathogens could help, but this again takes a number of years, and accepts the disadvantage of selection against only one environmental factor. If resistance is bought by the reduction of genetic variability or, even worse, a focus on few, or one, single genotype, then the parasite can again adapt to the host – in this case, a Pyrrhic victory is won. Beside this, there still exists the risk that other pathogens may occur which were unknown until then, and which easily could attack the tree population.

16.4 Genetic diversity as an element of design and planning in urban spaces

In order to obtain sustainable and robust urban tree populations, some authors suggest a high diversity of species and genera (e.g., Duhme and Pauleit, 2000; Raupp *et al.*, 2006; Bassuk *et al.*, 2009; Sjöman *et al.*, 2012). Thus, outbreaks of diseases caused by pathogens and insects should mainly be avoided or, at least, reduced. Barker (1975) was one of the first to suggest the use of a broader range of species. He suggested the guideline that no species should be represented by more than 5% of the total urban tree population. Other authors recommend values of around 10–15% (Grey and Deneke, 1986; Smiley *et al.*, 1986; Miller and Miller, 1991). Moll (1989) defines the specifications, insofar that he suggests 5% as the upper limit for a species, and not more than 10% of a genus. Santamour (1990) even goes a step further, and recommends benchmarks of 10% for a species, 20% for a genus and a maximal limit of 30% for a family.

Unfortunately, these authors neither give information on the area the data refer to, nor do they differentiate between different urban structures (e.g., larger park-like areas, smaller green spaces, city centre and peri-urban woodlands, street environment, etc.). Nevertheless, these recommendations, which are probably based more on experience than on scientific investigations, can serve as a first planning guide for increasing diversity of urban green. Beyond this aspect, intraspecific genetic variation is almost neglected. Thus, it remains an open question, to which extent comparable results can be gained by an appropriate genetic multiplicity within the single species, if a positive effect considering the guidelines can be significantly improved by increasing intraspecific genetic variation.

Information about the current state of diversity in cities is published in a number of articles. The overview of Sjöman *et al.* (2012) summarizes the results of a survey performed in ten north European cities. In this study, he identifies only one city out of ten which fulfils the requirement of maximal 10% of one species in the total tree population. Helsinki is an extreme example, with *Tilia x europaea* comprising a proportion of 44.7% of the total stock of trees. The genera *Acer* and *Tilia* together represent more than 57%. A comparable situation, with few genera dominating the total population, is present in another four cities (Sjöman *et al.*, 2012). It is also reported that the species diversity in parks is higher than that of street environments. A reason for this result may be the selection criteria "homogeneity" for street environments.

Although the situation in mid and southern Europe seems to be better, due to more favorable climatic conditions, the urban tree population in large cities generally consists of perhaps three, up to five, genera which represent 50–70% of the total street trees (Pauleit, 2002). The most common genera are *Tilia, Acer, Quercus, Aesculus,* and *Fraxinus.* For tropical conditions, a study by Pedlowski *et al.* (2002) carried out in Campos dos Goytacazes (a city in the north of Rio de Janeiro State of Brazil) showed that some species are extremely favored for planting in public areas due to factors such as their fast growth. In this case, *Caesalpinia peltophoroides* is the predominant species, ranging from 5–71% of the number of trees in the investigated neighborhoods. The situation seems to be similar in certain aspects to other Brazilian cities, where *Caesalpinia peltophoroides* is also planted in a large scale (80% of the trees in Rio Claro in the Brazilian state of São Paulo – Calvalheiro, 1991).

However, as previously mentioned, in the context of these studies, only the number of species or genera was in the focus of the investigations, not the intraspecific variability within the species.

How can the diversity aspect be involved in measures of urban tree management?

As the "evolution" of our cities and residential areas creates new sites and unpredictable environmental conditions, structures of high diversity should be considered if tree establishments are planned. In this way, the basis for a certain degree of genetic adaptability can be created. This can be achieved by the combination of plants belonging to different species or genera, but also by the conscious choice of different genotypes within a species (Krabel and Herzog, 2008).

In order to prevent calamities and diseases, plant material should be chosen in such a way that complexes of various plant ages are created. In areas where, due to aesthetic reasons, homogeneity in the age of the plant material is required, trees originating from seedlings (different genotypes) can be chosen (Santamour, 1990). If, for specific reasons, areas of genetically identical plant material are required, these complexes could be interrupted by groups of plants of other species or genera, or by genetically heterogeneous material from seedlings. An interruption of homogenous plant groups has to be regarded as prophylaxis against the spread of harmful parasites.

Existing structures which are planned to be completed should be completed with genetically diverse plant material and, preferably, with plants belonging to different batches and clones. If possible, small groups should at least be mixed with different species.

As a guideline, we suggest avoiding using genetically homogenous plant material wherever it is possible to do so.

16.5 Conclusions

Whether the suggested guidelines could avoid calamities and damage by environmental situations, especially in the context of climate change, has not been scientifically investigated up to now. Experience from agriculture and forestry, however, leads to the conclusion that an increased genetic diversity within plant/tree populations will have positive effects on the vitality and lifetime of urban tree populations. Besides the restrictions which are imposed by site conditions, concepts of urban planning and management should generally integrate guidelines to achieve a minimum of diversity. This does not mean that "new", rather than autochthones tree species, should be used in an unregulated way, because the introduction of "new" genera and species holds the well-known risk that, often, little information is available about the behavior of the species in a new environment (see also Chapter 14, potentially invasive species), and the impact on our ecosystems is often larger than expected.

In this context, especially in landscape planning and forestry, increased attention is paid to using autochthones plant material. In fact, the intraspecific resources of native species should be exploited more intensively. Practically, this means that, among the genotypes of a species existing in nature, genotypes can be selected for future urban tree breeding that are adapted to the environmental conditions which are typical for urban environments (e.g., plants growing on rather dry sites, sites with increased salt contamination, sites with increased air pollution).

A practical guideline should be based on scientific investigations, which requires intensive cooperation in research and practice. The same is true for the implementation of these concepts. The success will surely depend on how intensively breeders, planners and nurseries will cooperate.

References

Barker, P. (1975). Ordinance control of street trees. *Journal of Arboriculture* **1**, 212–215.

Bassuk, N., Deanna, F.C., Marranca, B.Z. and Barb, N. (2009). *Recommended urban trees: Site assessment and tree selection for stress tolerance*. Urban Horticulture Institute, Cornell University, Ithaca, New York.

Cavalheiro, F. (1991). Urbanição e alterações ambientais. In: Tau, S.M. (ed). *Análise Ambiental: Uma Visão Multidisciplinar*, Editora UNESP, São Paulo, Brazil.

Duhme, F. and Pauleit, S. (2000). The dendrofloristic richness of SE-Europe, a phenomenal treasure for urban plantings. *Mitteilungen aus der Biologischen Bundesanstalt für Land- und Forstwirtschaft Berlin-Dahlem* **370**, 23–39.

Gaston K.J. (1996). *Biodiversity: A Biology of Numbers and Difference*. Blackwell Science, Oxford, 396 p.

Grey, G.W. and Deneke, F.J. (1986). *Urban Forestry*, 2nd edition. Wiley & Sons, New York.

Krabel D. (2011). Anpassungspotentiale forstwirtschaftlich relevanter Baumarten. *Allgemeine Forst Zeitschrift – Der Wald* **71**(11), 8–9.

Krabel D. and Herzog S. (2008). Genetische Variabilität als Voraussetzung für den Erhalt von Stadtbäumen. *Forst und Holz* **63**(3), 23–25.

Miller, R.H. and Miller, R.W. (1991). Planting survival of selected street tree taxa. *Journal of Arboriculture* **17**, 185–191.

Moll, G. (1989). Improving the health of the urban forest. In: Moll, S. and Ebenreck, S. (eds). *A resource guide for urban community forests*, pp. 119–130. Island Press, Washington.

Parson, T.J., Bradshaw Jr., H.D. and Gordon, M.D. (1989). Systemic accumulation of specific mRNAs in response to wounding in poplar trees. *Proceedings of the National Academy of Sciences of the United States of America* **86**, 83–96.

Pauleit, S., Jones N., Garcia-Martin, G., *et al.* (2002). Tree establishment practice in towns and cities – Result from a European survey. *Urban for Urban Green* **1**(2), 83–96.

Pedlowski, M.A., Corabi Adell, J.J. and Heynen, N.C. (2002). Urban forest and environmental inequality in Campos dos Goytacazes, Rio de Janeiro, Brazil. *Urban Ecosystems* **6**, 9–20.

Raupp, M., Cumming, M. and Raupp, E.C. (2006). Street tree diversity in eastern North America and its potential for tree loss to exotic borers. *Arboriculture & Urban Forestry* **32**(6), 297–304.

Roloff, A. and Bärtels, A. (2014). *Flora der Gehölze – Bestimmung, Eigenschaften, Verwendung*, 4th edition. Ulmer Verlag, Stuttgart.

Santamour, F.S. (1990). Trees for urban plantings: diversity, uniformity and common sense. In: Proceedings of the 7th Conference of the Metropolitan Tree Improvement Alliance, Vol **7**, pp. 57–65.

Schlichting, C.D. (2002). Phenotypic plasticity in plants. *Plant Species Biology* **17**, 85–88.

Sjöman, H., Östberg, J. and Bühler, O. (2012). Diversity and distribution of the urban tree population in ten major cities. *Urban Forestry & Urban Greening* **11**, 31–39.

Smiley, E.T., Kielbaso, J.J. and Proffer, T.J. (1986). Maple disease epidemic in southeastern Michigan. *Journal of Arboriculture* **12**(5), 126–128.

Stettler, R.F., Bradshaw Jr., H.D. (1994). The choice of genetic material for mechanistic studies of adaptation in forest trees. *Tree Physiology* **14**, 781–796.

United States, Congress, Office of Technology Assessment (1987). *Technologies to maintain biological diversity*. Congress of the U.S. Office of Technology Assessment, Washington, D.C.

CHAPTER 17

Governance in urban forestry

Jürgen Pretzsch

Technische Universität Dresden, Tharandt, Germany

17.1 Introduction: challenges and need for action

Increasing urbanization worldwide leads to new challenges regarding the organization of individual lives, as well as urban society as a whole. At present, more than 50% of the world's population live in urban agglomerations (UN, 2014). The level of urbanization and the character of urban agglomerations is diverse, determined by ecological, cultural and economic factors. The proportion of urbanization over rural population is still low in Africa and some other countries of the Global South, but it is increasing drastically, with even more challenging consequences than in industrial countries (UN, 2014).

The demand for urban forest services is rather different in relation to the gross national product or the livelihood index. In industrial countries, the demand pattern is characterized by a post-materialist value system (Weber and Mann, 1997). It focuses on recreation and sports, balancing workdays by nature experience, stress reduction and biological food production. In countries of the Global South, a major goal of urban forestry is still basic needs satisfaction, related to the provision of food, energy and water, as well as housing and medical care (Kuchelmeister, 2000). Additionally, an increasing level of environmental services, such as green spaces for recreation, healing and health, as well as environmental education, is in demand.

The access to urban green is becoming more and more unequal (Heynen *et al.*, 2006). The population capacity of the planet Earth is not primarily limited by physical factors like food production, but options and restriction of social organization (Heilig, 1996). Increasing conflicts in urban forestry are related to security problems, land access or exclusion, benefit sharing in urban forestry services, and products and challenges related to survival in general.

However, any generalization is difficult, and only a synoptic overview is given in this chapter, without any claim to presenting transferable problem solutions.

17.2 Objectives and definitions

17.2.1 Objectives

Urban forestry is profoundly interrelated to urban development, as well as with the general socio-economic and cultural development patterns at upper spatial scales, which is why a site- and context-specific analysis is necessary. The objective of this contribution is

Urban Tree Management: For the Sustainable Development of Green Cities, First Edition. Edited by Andreas Roloff.
© 2016 John Wiley & Sons, Ltd. Published 2016 by John Wiley & Sons, Ltd.

to synthesize the history, present shape and expected future development of governance models in urban forestry. These are relevant for decision-making and organizational development. The integration of the governance models in an overall co-evolution approach is a further objective.

17.2.2 Definitions

Governance expresses decision making and acting in a multiple-stakeholder constellation. Stoker (1998, 19) defines governance as a *"complex set of institutions and actors that are drawn from but also beyond government"*. This is opposed to the simple *government* organization; it does not simply follow the hierarchical structure of decision-making and action in public authorities. Depending on the people and institutional settings involved, organizational structures have developed and are undergoing permanent change.

Related to *institutions*, the definition of North (1991, 97) is applied: *"Institutions are the humanly devised constraints that structure political, economic and social interaction. They consist of both informal constraints (sanctions, taboos, customs, traditions, and codes of conduct), and formal rules (constitutions, laws, rights)."* The resulting organizational units are composed of all participants directly involved in decision-making and execution of urban forestry activities. This is not necessarily related to a formal mandate. As in neighborhood groups, informal rules and actor interrelations frequently play an essential role and shape the organization of action.

There is a trend towards more *participation* of diverse members of the civil society in decision-making and action. Arnstein (1969, 217) differentiates the following *"eight rungs on a ladder of citizen participation"*: manipulation; therapy; informing; consultation; placation; partnership; delegated power; and citizen control. This scale is helpful for the diagnosis of stakeholder participation in urban forestry.

The discussion of *urban* forestry includes *peri-urban* forests and trees. The interrelation between centers of cities and their surroundings plays an increasing role in expanding global urbanization. Of relevance is the provision of a multitude of demanded goods and services from peri-urban areas for employees and residents of the city centre (Allen, 2003; Pretzsch, 2014).

The *livelihood approach* permits a differentiated observation and analysis of resident households, based on the five dimensions of natural, physical, human, social and financial capital (Carney, 2003). Here, special emphasis is put on *social capital* (see section 17.3.5).

17.3 Diagnosis and conceptual framework

17.3.1 Socio-ecological co-evolution model for urban forestry

The action and research field of urban forestry governance is best structured in a socio-ecological co-evolution model (see Figure 17.1). This conceptual model was initially elaborated to explain rural development patterns (Pretzsch, 2014). More recently, social-ecological models have increasingly been applied to analyze respective interrelations in urban forestry (Mincey *et al.*, 2013). The co-evolution model is composed of the natural system (I), the social system (II) and the interface (III), which is focused on governance options in this publication.

The ecological system (I) represents a large variety of green spaces. These are mostly planted units, including street trees, parks or home gardens. It may also be composed of succession vegetation, such as in abandoned industrial installations, along railway lines or in conservation areas. The constituent elements of the social system (III) are the

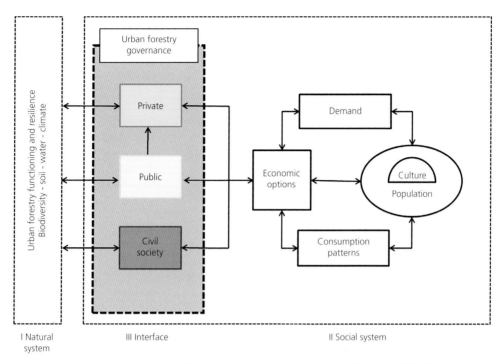

Figure 17.1 Socio-ecological co-evolution model (derived from Hurni and Messerli, 1981; Norgaard, 1994; Berkes *et al.*, 1998; Pretzsch, 2014).

demand for urban forestry products and services, the respective segment of the population, their consumption pattern and the economic options. The interface (III) represents governance options, which involve institutional and organizational patterns. The differentiation in public, private and civil society mechanisms underlines the variety and complexity of interactions.

17.3.2 Historical development of urban forestry governance

A historical revision of urban forestry governance models will facilitate understanding of the present situation. A respective socio-economic diagnosis is composed of a historical assessment, as well as a comparison of the outcome of this assessment with an afterwards developed set of normative targets. Here, the normative targets are based on the concept of a *Green economy*, which is well documented by UNEP (2011).

Urban forestry governance is rooted in the traditions of public administration, accompanied by bureaucratic models, which were based on control and sanctions and had the character of local government interventions. In parallel, various urban green spaces were private properties, in which certain management rules had to be followed. In Europe, urban green spaces, such as parks, were traditionally managed by the local administration, or by the respective private owners and residents. This administrative management system has the advantage of direct responsibility, and a high level of qualification of communal employees. It was a fundamental aspect of many sophisticated urban parks, tree-lined walks and gardens in towns. Even under extraordinary pressure, such as in times of war, food shortage or other crises, the urban green was kept under control. Tree-cutting for firewood production took place just sporadically.

Figure 17.2 Urban forestry has to adapt to complex urban and peri-urban structures (Photograph by Robert Schreier, Project Megacities Recife. Reproduced with permission).

Following the growth and diverse spatial arrangement of urban areas (see Figure 17.2), the demand for green spaces and urban forests has increased and diversified. The implementation of urban forestry became more difficult and complex, and the multiple disadvantages of purely administrative management became obvious. The increasing cost could no longer be covered by communal budgets. Because of the low participation in decision-making, there was hardly any motivation for civil society to engage in urban forestry. Often, the administration followed a relatively "traditional" technical focus, not immediately taking in account new, upcoming needs of the population. Rapid urbanization, together with the changing demands of society, required new forms of governance in urban green space management. Although, in the past, communal and state organizations were most frequently in charge of all steps of the management cycle in urban forestry, additional participants now needed to be involved. In the United States of America, a drastic shift to the involvement of residents and of the private sector has taken place since the middle of the last century.

17.3.3 Increasing complexity and paradigm change

Many changes can be explained with the rise of a new paradigm at the end of the 20th century. Paradigms are defined as "the entire constellation of beliefs, values, techniques, shared by members of a given community" (Kuhn, 1962, 175). They follow four characteristics:

1 They are rooted in theories.
2 They have an impact on policies.

3 They are articulated by a clearly determined group of scholars.

4 They exert influence in practice.

The new theoretical fundaments are related to multiple stakeholder involvement in decision-making and collective choices. Besides the communal administration, a number of other participant groups take over responsibilities and decisions in urban green space management.

Among these are interest groups, non-governmental organizations and a number of new upcoming contributors from the private sector, including shops, industries and small landowners, as well as public entities like schools, hospitals, and churches. Their influence in decision-making and urban policy making is increasing. Scientific disciplines have reacted with a number of governance models and theories, which explain multi-stakeholder decision-making and their importance for green space management (Jones *et al.*, 2005). Finally the new paradigm has an important influence in practice. More participants are involved in urban forestry and, in recent years, the subject matter has become important in the discourse on environmental management, as well as an element of the green economy concept.

17.3.4 Stakeholder analysis and differentiation in participant groups

A stakeholder analysis facilitates the selection of all relevant stakeholders for an urban forestry initiative, based on their interest, importance and relevance for decision-making. These insights permit differentiation at a horizontal level, for example between different income groups in a city. Additionally, hierarchical interrelations are documented. This focuses on multi-level governance mechanisms – for example, the relationship between the city administration and residents, neighborhood groups or business units (Grimble and Wellard, 1997).

Generally, stakeholders can be clustered into three groups (see Figure 17.3): the public sector, including the administration; the private sector, composed of individual residents, shops and other business units as industries; and civil society stakeholders and organizations, which may have different levels of institutionalization and membership

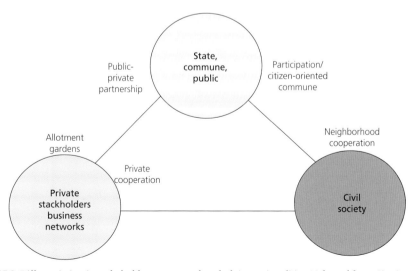

Figure 17.3 Differentiation in stakeholder groups and underlying rationalities (Adapted from Sinning 2006, 87).

obligations. Examples are loosely linked neighborhood groups, with informal roles on the one hand, and legally inscribed associations on the other hand.

The three groups follow different rationalities. The behavior of the players in the public sector is explained by the *public choice* theory, following the rationality of the administration. In democratic states, they depend, at least to some extent, on formal administrative rules, clear hierarchies, and public elections of a politically legitimized head. The private sector follows the market rationality. Liberal economists assume, in the underlying *rational choice* principles, that maximization of individual utilities contributes best to urban development. Often, far-reaching privatization of land and other properties is seen as a necessary precondition for successful development. In urban forestry, this needs to be critically questioned. In the countries of the global south, especially, privatization may lead to the exclusion of parts of society from access to urban green spaces and the offered services.

A third important path is the management of green spaces by civil society institutions. This model is rooted in the concept of *collective action*, and is based on group activity. Decisions are taken by groups, and solidarity or additional benefits from group interaction may play an important role. In countries of the global south, especially, collective action-based initiatives play an increasing role. Green spaces on free-access communal land provide food, construction materials and other livelihood means. Cooperation and coalitions between the three stakeholder groups are common. They are indicated in the triangle in Figure 17.3.

17.3.5 Assessment by the livelihood framework

On the level of residents, the livelihood approach permits assessment of the needs of the involved residents, as well as scenarios and development strategies on urban forestry development (Carney, 2003). Related to urban green spaces and their use, the respective five capitals allow checking development options, together with the residents. Natural capital defines the physical conditions of tree planting, physical capital the existence of respective tool and plants, human capital the education and knowledge, and financial capital the investments and cash flow, related to urban trees and parks. Social capital, which is related to networks, trust and values, is of special relevance in collective action strategies (see Figure 17.4). Many of the outlined components of social capital need further investigation

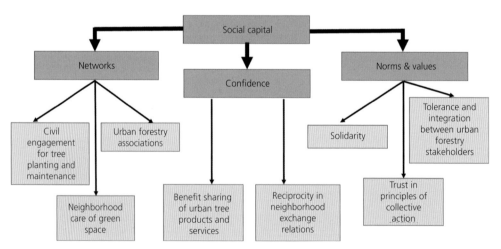

Figure 17.4 Components of social capital (Adapted from Van Deth 2003, 83).

in their application to urban forestry. In multi-stakeholder management concepts, different types of capital may be provided by different players, integrating it into a common development strategy.

It can be assumed that, in parallel to urbanization in general, the importance of urban green spaces and urban forestry will increase rapidly. The high complexity of the social and economic structure implies the need for diversification of governance approaches, which permit a proper implementation and maintenance of the green spaces.

17.4 Governance models for urban forestry

17.4.1 Introduction to urban forestry governance models

The paradigm change has led to the implementation of a multitude of multi-stakeholder governance and organizational models. So far, they have been little investigated in detail, and scientific documentation is still lacking. Besides the increasing importance of urban forestry, there is a boom in urban agriculture system development. This is mostly focused on vegetable production, and trees play a minor role, because they require long-term disposition and investment. The outlined models represent just a selection of possible options (Clark *et al.*, 1997).

17.4.2 Public administration: changing functions and diversification

Besides executing traditional tasks such as tree planting and maintenance, new activity fields have developed in urban administrations. With the abovementioned paradigm change, much more coordination is necessary between different stakeholder groups. In most of the governance models, public administration is at least a minor partner, and is responsible for the overall integration in urban planning. In countries of the global south, especially, an additional task is balancing the offer of urban forestry services in the whole area of a city, integrating urban districts with social problems caused by unemployment, suppression of ethnic minorities or poverty. An increasing number of investigations prove that poorer populations of city inhabitants, in particular, have only a limited access to green spaces (Heynen *et al.*, 2006).

The system of public administration of urban forestry and green spaces has reached its limits in many towns, and various stakeholder groups want to participate in decision-making. Budgets are increasingly restrictive, and just "obligatory" tasks are funded by public means. The financial situation of most public entities is difficult, as international conditions and austerity programs cut down public expenditure. The voluntary community expenditure is especially being reduced. Mostly, public expenditure for urban forestry does not follow a legal obligation, and the availability of public funds is decreasing. As a consequence, the public administration depends more and more on coalitions with other stakeholder groups including interest groups or non-governmental organizations. A simple way is the outsourcing of activities to the private sector – for example, green space planning done by private services of landscape architects, park control by a security service enterprise, and planting and maintenance activities by gardening firms or green space-based event management.

Besides outsourcing, activities are increasingly executed by the public administration in partnership with others. Examples include the establishment and maintenance of tree-lined walks in cooperation with residents. The administration gives advice to the residents about species selection or planting technique, and provides the plants in some cases.

The residents, or resident groups, take over planting and maintaining the trees. In tropical countries, watering is an essential and time-consuming task. If the cooperation is formalized, public-private partnership models are frequently applied.

17.4.3 Public-private partnerships

Many former tasks of the municipality have been taken over by public-private partnership (PPP) agreements. Financially strong partners include large business units or industries, which want to contribute to making the environment around their property attractive. The obligations of both partners are regulated in agreements and contracts. Most PPP agreements permit a certain planning security to the municipalities. This type of cooperation is increasingly implemented around the rapidly increasing number of shopping centers, which are often situated in peri-urban areas, in large parking grounds and around business centers.

17.4.4 Governance based on private urban forestry

Private stakeholders contribute significantly to maintaining green spaces and urban forests in numerous towns. In many countries, private residents must coordinate their planting and removal of urban trees with the public municipality or, increasingly, with civil society organizations. Governance strategies differ between resident groups. Large property owners often dispose of extensive gardens, which are not accessible for the public. Exclusion is increasingly a feature seen in countries of the global south; because of security problems, fences around residential areas have become more and more common. A typical example are the fenced middle class barriers (condominios) in Brazil (see Figure 17.5).

Meanwhile, for residents in poor areas, development options for urban green and tree planting are limited (Figure 17.6). Many rural residents with home gardens in peri-urban areas experience strong pressures and, because of dramatically rising land prices, land speculation and increasing criminality, they often give up their farms and sell their land.

A special governance model deals with agreements between private stakeholders. Private residents may be involved in agreements with shopping centers or industrial installations, which take over the cost of green space development, establishment and maintenance in the surrounding area. Benefits for the large private units are good publicity and image, and attractive condition for the customers. Meanwhile, the private landowners profit from a subsidized improvement of the area, which even may have an effect on property values.

17.4.5 Donations

An increasing number of non-profit organizations pursue the target of city greening, with budgets based on donations. In the United States of America and Canada, especially, there is a real boom in the formation of respective non-governmental organizations. Besides tree planting, environmental education is offered by civil society organizations. Frequently, didactical materials are reproduced and distributed on environmental education (see Chapter 20).

17.4.6 Allotment gardens

Allotment gardens represent a typical governance system between communal land owners and private urban residents, who want to enjoy growing horticultural plants and harvesting the produce from them, including fruits from trees. Today, these gardens play

Figure 17.5 Private fenced gardens in the condominios; upper class residential areas (Photograph by Robert Schreier, Project Megacities Recife. Reproduced with permission).

Figure 17.6 Public spaces without much option for tree cover (Photograph by Robert Schreier, Project Megacities Recife. Reproduced with permission).

an important role for outdoor recreation for families living in cities with little green space. Access to an allotment garden requires membership of an association, which usually has relatively strict rules about how to manage the land (see Chapter 18). This urban garden governance model has transferred to countries of the global south, but its adoption has remained rather limited. Availability of land is restricted, and the specific criteria for their creation is often is not made public. Disconnection between the location of housing and gardens may lead to additional problems, such as control, watering and other maintenance.

17.4.7 Neighborhood groups and collective gardening

Neighborhood initiatives are often rooted in collective action measures. Residents organize themselves informally in interest groups and look after the green space in their district, often combined with caring for playing grounds, safety measures or even common childcare. In recent times, this has been extended to gardens which are shared by members of one group. Of importance are rules for the distribution of tasks and a minimum of responsibilities. Increasingly, the production of agricultural goods takes place on common land, which also requires clear regulations between the inhabitants who are involved (Ostrom, 1990).

17.5 Lessons learned for the future development of urban forestry

17.5.1 Paradigm change

The paradigm change from public urban forestry administration to multi-stakeholder constellations is advancing with different speed, depending on the specific situation in towns and countries. It is obvious that, especially in the USA and Canada, the shift towards the involvement of the private sector is much more advanced than in Europe and other continents. In a large number of towns, specific urban forestry master plans have been elaborated. Public-private partnership models and sponsoring have high priority. Meanwhile, the involvement of various groups in collective action is still under expansion. Recently, political ecology studies are expanding towards urban areas, and imbalances in stakeholder, benefit distribution and power structures are being investigated (Sandberg *et al.*, 2015). Figure 17.7 demonstrates the sequence in stakeholder involvement.

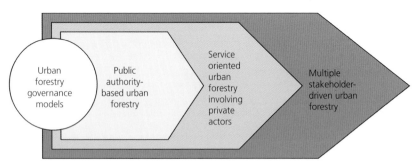

Figure 17.7 Sequence from public to multi-stakeholder driven urban forestry (Adapted from Sinning 2006, 88).

17.5.2 Chances and limits of collective action in urban forestry

With the increasing complexity of the functions of urban spaces, the multitude of demands of the residents and the application of new organizational forms, collective action-based governance models are on the rise (Peet and Watts, 1996). The traditional neighborhood model is complemented by shared gardens, related food cooperatives, slum greening initiatives and green kindergarten projects. In Brazil, Singer (2002) has developed viable concepts for the development of slum areas in megacities, based on solidarity economy. This means that the residents take action with self-help mechanisms, including gardening and tree planting, even propagating the development of universities in slum districts.

All of these initiatives require a basic set of rules to channel the individual activities. Ostrom (1990) outlines nine principles which are applicable for the management of free access urban forestry areas: clearly defined boundaries; congruence; collective action arrangements; monitoring; graduated sanctions; conflict resolution mechanisms; minimal recognition of rights to organize; and nested enterprises. There is a strong discussion on the limits of self-governance in urban areas (Davies, 2002).

17.5.3 Exclusion and conflict management

Many community green spaces are there for the public good, so direct exclusion is not possible and "free riders" may misuse the free access. This can vary with location and average income, as well as other social and cultural parameters such as ethnicity, caste and education. In tropical countries, in particular, the poor sections of the urban population are indirectly excluded from the use of green space and, often, there is hardly any urban forest at their disposition. This may have a strong effect on health, child education, moral stability and livelihood in general. Together with increasing complexity in the organization of stakeholders who are involved in urban forestry, conflicts are also on the increase. As a consequence, conflict management strategies have to be elaborated proactively as part of governance strategies, with the objective of motivating all players to participate in urban forestry.

Conflicts frequently occur in all stages of urban forestry planning and implementation. The more complex the stakeholder constellations, the more difficult is proper planning and the integration of all relevant participants. A stakeholder analysis plays an essential role. The conflicts often occur because perception and attitudes differ between the involved stakeholder groups (see Figure 17.8). Attitudes are a complex mix of cognitive, emotional and experience-based perception, and often are not clearly visible. Possible reasons for conflicts are a generally aggressive attitude in the urban environment, due to the dramatic alienation of urban residents from nature, and demands for urban forestry services not being satisfied. There is a need to proactively foresee these conflicts by installing conflict management systems. This permits dysfunction to be perceived and mediated at an early stage, before conflicts enter into an elevated stage of aggression (Glasl, 1999).

17.5.4 Adaptive management

The increasing complexity and the necessarily non-linear implementation of urban forestry initiatives require an iterative procedure. This is due to multiple stakeholder involvement, as well as to the learning character of new initiatives and steps. This is of special relevance in large cities with unequal income distribution and social tensions, where integrative urban forestry planning is increasingly challenging. The implementation

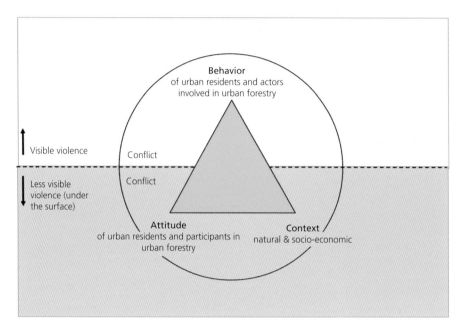

Figure 17.8 Conceptual approach to identify conflicts in urban forestry (Adapted from Specht 2008, 7).

of adaptive management requires innovative staff and permanent training (Williams, 2011; Shindler and Cheek, 1999).

17.5.5 Forthcoming steps in practice and research

Increasing urbanization leads to a high complexity of various governance structures. This requires more research in urban forestry governance, and the elaboration of a respective theory framework. In a first step, the analysis of case studies permits advances in the understanding of innovative governance models. Lawrence *et al.* (2013) suggest a standardized procedure, and present a criteria catalogue to advance with case study research. Intensive cooperation between urban forestry management units and research institutions is necessary for this.

The comparative analysis of case studies permits the categorization of urban forest governance models, and the elaboration of respective hypotheses. In a further step, this leads to theory building, which permits synthesis of the case study experiences to make the attained knowledge transferable. Challenging new trends indicate the development of new green cities, which integrate food production, recreation, health and high levels of employment (like Singapore). Essential to this is integration in a holistic concept of a green economy. This may only be implemented with strong inter-sector and interdisciplinary cooperation.

17.6 Conclusions

Conclusions are best linked to the socio-ecological co-evolution model (see Figure 17.9). Further development of the ecological system (I) is necessary. At present, techniques of urban agriculture are under intensive investigation, and the integration of trees is

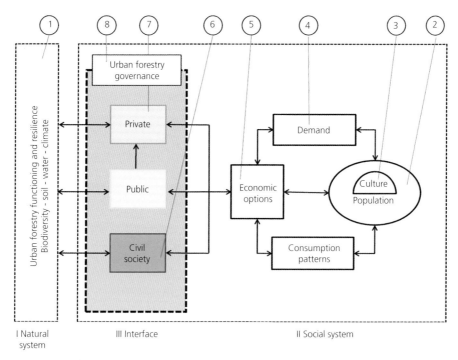

Figure 17.9 Co-evolution model with the indication of priority subject in urban forestry governance (Derived from Hurni and Messerli, 1981; Norgaard, 1994; Berkes *et al.*, 1998; Pretzsch, 2014).

important for the stabilization of the ecological systems as well as for the additional provision of products and services.

Climate change adaptation (1 in Figure 17.9) will play an important role in the interrelation between anthropological causes and necessary adaptation. The social system (II) is extremely dynamic, and forecast of its parameters is difficult. It is obvious that the urban population will increase drastically in the next decades (2). Globalization might, as a counteraction, lead to an empowerment and a differentiation of ethnic urban cultures (3), which leads to a large variation in demand patterns, depending on ethnic groups (4). Different attitudes towards urban forestry are also partly dependent on the economic status (5).

With the increase of urbanization, the demand pattern of urban residents will change. If the present nearly global process of increasing market orientation goes on, a rising part of urban residents will be indirectly excluded from urban forestry access, products and services (5). To avoid this, urban forestry governance systems (III) need to be further developed. It is obvious that, besides the town administration, special emphasis has to be put on collective action mechanisms (6), which permit the involvement of a large diversity of stakeholders. However, the respective governance systems are extremely complex and difficult to institutionalize.

The free rider problem has special relevance in towns which are, compared with rural areas, even more characterized by low transparency. The integration of poor residents of towns into collective action-driven activities may create better understanding of green spaces and motivate the planting and maintenance of trees.

Further privatization (7) of urban forestry facilities, including parks and gardens, may lead to more exclusion and conflicts, with leakage effects in other green spaces in a town. The implementation of complex governance systems requires the proactive installation of conflict management mechanisms (8).

The co-evolution model demonstrates that the elaboration, testing and implementation of urban forestry governance strategies requires an interdisciplinary focus. There is a strong interrelation between ecological and technical aspects, socio-economic development and the respective governance strategies. Teamwork is necessary in practice, as well as in the sciences. In the megacities of the future, experts from a broad mix of different disciplines will be needed to manage urban forestry, involving arboriculture, ecology, sociology, economy, landscape architecture, urban planning and education, as well as policy and anthropology. Necessarily management and governance strategies have to follow an adaptive approach, because many patterns are changing, and learning is necessary in all stages.

References

Allen, A. (2003). Environmental planning and management of the peri-urban interface; perspectives on an emerging field. *Environment & Urbanization* **15**(1), 135–148.

Agrawal, A. and Ostrom, E. (2001). Collective action, property rights and the devolution of forest and protected area management. In: Meinzen-Dick, R., Knox, A. and Di Gregorio, M., (eds). *Collective action, property rights and devolution of natural resource management. Exchange of knowledge and implications for policy*. Proceedings of the International Conference held from 21–25 June in Puerto Azul, The Philippines. DSE/ZEL Feldafing, pp. 75–109.

Arnstein, S.R. (1969). A ladder of citizen participation. *Journal of the American Institute of Planners* **35**, 216–224.

Bell, S. (2003). Contested views of freedom and control: Children teenagers and urban fringe woodlands in environmental pollution. *Urban Forestry & Urban Greening* **2**, 87–100.

Berkes, F., Folke, C. and Colding, J. (1998). *Linking social and ecological systems*. Cambridge University Press, Cambridge.

Buchy, M. and Hovermann, S. (2000). Understanding public participation in forest planning: a review. *Forest Policy and Economics* **1**, 15–25.

Carney, D. (2003). *Sustainable Livelihoods Approaches: Progress and Possibilities for Change*. DFID, Toronto.

Carter, J. (1995). *The potential of urban forestry in developing countries: A concept paper*. FAO, Rome. Available under http://www.fao/docrep/oo5/t1680e00.htm [accessed March 2015].

Clark, J.R., Matheny, N.P., Cross, G. and Wake, V. (1997). A model of urban forest sustainability. *Journal of Aboriculture* **23**(1), 17–29.

Davies, J.S. (2002). The governance of urban regeneration: A critique of the 'Governing without government' thesis. *Public Administration* **80**(2), 301–322. Available at: http//ssrn.com/abstracts=1813749 [accessed March 2015].

Glasl, F. (1999). *Konfliktmanagement. Ein Handbuch für Führungskräfte, Beraterinnen und Berater*. Haupt, Bern.

Grimble, R. and Wellard, K. (1997). Stakeholder methodologies in natural resource management: a review of principles, contexts, experiences and opportunities. *Agricultural Systems* **55**(2), 173–193.

Heilig, G.K. (1996). How many people can be fed on earth? In: Lutz, W. (ed). *The future population of the world; what can we assume today?* pp. 196–249. Earthscan, London. Revised and updated edition.

Heynen, N., Perkins, H.A. and Roy, P. (2006). The political ecology of uneven urban green space. The impact of political economy on race and ethnicity in producing environmental inequality in Milwaukee. *Urban Affairs Review* **42**(1), 3–25.

Hurni, H. and Messerli, B. (1981). Mountain research for conservation and development in Simen, Ethiopia. *Mountain Research Development* **1**(1), 49–54.

Jones, N., Collins, K., Vaughan, J., Benedikz, T. and Brosnan, J. (2005). The role of Partnerships in Urban Forestry. In: Konijnendijk, C., Nilsson, K., Ranrup, T. and Schipperijn, J. (eds). *Urban Forests and Trees. A Reference Book*, pp. 187–205. Springer, Heidelberg.

Kuchelmeister, G. (2000). Trees for the urban millennium: urban forestry update. *Unasylva (FAO)* **200**, 49–55.

Kuhn, T.S. (1962). *The structure of scientific revolutions*. University of Chicago Press, Chicago.

Lawrence, A., De Vrese, R., Johnston, M. and Konijnendijk (2013). Urban forest governance: towards a framework for comparing approaches. *Urban Forestry & Urban Greening* **12**(4), 464–473.

Mehta, D. (1998). *Urban Governance: Lessons from Best Practices in Asia*. UMP-Asia Occasional Paper 40. http://gdrc.org/u-gov/good-governance.html [accessed March 2015].

Miller, R.W. (1988). *Urban Forestry. Planning and Managing Urban Greenspaces*. Prentice Hall, New Jersey.

Mincey, S.K., Hutten, M., Fischer, B., Evans, T., Stewart, S. and Vogt, J.M. (2013). Structuring institutional analysis for urban ecosystems: A key to sustainable urban forest development. *Urban Ecosystems* **16**(3), 553–571. DOI: 10.1007/s11252-013-0286-3.

Newman, J. (2001). *Modernising Governance*. Sage, London.

Norgaard, R.B. (1994). *Development betrayed the end of progress and a coevolutionary revisioning of the future*. Routledge, London and New York.

North, D.C. (1991). Institutions. *Journal of Economic Perspectives* **5**(1), 97–112.

Ostrom, E. (1990). *Governing the Commons: the evolution of institutions for collective action*. Cambridge University Press, California, New York.

Peet, R. and Watts, M. (1996). Liberation ecologies. Environment, development, social movements. Psychology Press, London/ New York.

Pretzsch, J. (2014). Paradigms of Tropical Forestry in Rural Development. In: Pretzsch, J., Darr, D., Uibrig, H. and Auch, E. (eds). *Forests and Rural Development*, pp. 7–49. Springer, Berlin/ Heidelberg.

Sandberg, A., Bardekjian, A. and Butt, S. (eds, 2015). *Urban forests, trees and greenspace. A political ecology perspective*. Earthscan, London and New York.

Shindler, B. and Cheek, K.A. (1999). Integrating citizens in adaptive management: A propositional analysis. *Conservation Ecology* **3**(1) (online). Available at: http://www.consecol.org/vol3/iss1/ art9/ [accessed March 2015].

Singer, P. (2002). *Introducao à Economia Solidária*. Fundacao Perseu Abramo, Sao Paulo.

Sinning, H. (2006). Urban Governance und Stadtentwicklung – Zur Rolle des Bürgers als aktiver Mitgestalter und Koproduzent. *VHW Forum Wohneigentum* **1**, 87–90.

Specht, I. (2008). *Conflict analysis. Practical tool to analyse conflict in order to prioritise and strategise conflict transformation programmes*. Available at: www.search4dev.nl/download/282454/114718.pdf [accessed March 2015].

Stoker, G. (1998). Governance as theory: five propositions. *International Social Science Journal* **50**(155), 17–28. Available at: catedras.fsoc.uba.ar/rusailh/Unidad 1/Stoker 2002, Governance as theory, five propositions.pdf [accessed March 2015].

UN (2014). *World Urbanization Prospects – The 2014 Revision Highlights*. United Nations, New York, NY.

UNEP (2011). *Green Economy report. Towards a green economy: pathways to sustainable development and poverty eradication*. Available at: www.unep.org/greeneconomy/ GreenEconomyReport, free release 2011 [accessed March 2015].

Uphoff, N. (1993). Grassroots organizations and NGOs in rural development: Opportunities with diminishing states and expanding markets. *World Development* **21**(4), 607–622.

Van Deth, J.W. (2003). Measuring social capital: orthodoxies and continuing controversies. International. *Journal of Social Research Methodology* **6**(1), 79–92.

Weber, N. and Mann, S. (1997). Der postmaterialistische Wertewandel und seine Bedeutung für die Forstwirtschaft. *Forstarchiv* **68**, 19–24.

Williams, B. (2011). Adaptive management of natural resources – framework and issues. *Journal of Environmental Management* **92**, 1346–1353.

CHAPTER 18

Allotment gardens and privately managed green space in urban environment

Eckhard Auch

Technische Universität Dresden, Tharandt, Germany

18.1 Introduction

Why talk about gardens in a book about urban trees? Since trees are an integral part of gardens, and gardens are an integral part of urban/peri-urban areas, it is worth taking a look at this element of urban greenery. From a gardener's point of view, dividing up the elements of a garden into trees and other plants goes against his concept of what a garden is (Fotschki, 2014). Although trees comprise the nexus of this chapter, with their important functions for providing shade, food, amenity, local climate, habitats, and so on (see Chapter 1), the units of management are, however, gardens. Here, focus is given to urban allotment gardens. Allotment gardeners prove their motivation month by month by paying rent, and it can therefore be assumed that they make more conscious decisions as to the functions their garden should fulfill, as opposed to gardens which simply constitute a by-product of housing.

18.2 Some definitions

18.2.1 Green space as urban soft infrastructure

Cities are characterized by a high density of constructed "hard infrastructure", but there are, nevertheless, still gaps, spaces or plots with more or less natural vegetation, parks, gardens, forests, farms, and so on. These vary from micro-sized to large, and they belong to a city's "soft infrastructure", meaning "those things that enhance the quality of life" (Bennis, 2006; and see Chapter 1). These diverse green areas and surfaces interact in many ways with each other, with other elements of their natural environment and with surrounding buildings, as well as with people (Carreiro and Zipperer, 2008). Modern urban "green" has conventionally been shaped by cultural aspects. Like the buildings, the green space has been developed with elements of selected, modified nature, designed according to the prevailing paradigms and mainstreams of architecture. Post-modern designs (responding to financial constraints in municipal budgets) increasingly include succession and wild/unmanaged nature in urban green space concepts.

Urban Tree Management: For the Sustainable Development of Green Cities, First Edition. Edited by Andreas Roloff.
© 2016 John Wiley & Sons, Ltd. Published 2016 by John Wiley & Sons, Ltd.

Zoning plans, neighborhood laws and jurisdiction on damages caused by trees delimit the role of the tree component in urban/peri-urban gardens. Land tenure is also an important factor. Usually, people are very reluctant to plant trees if they have no security of tenure. They fear the risk of losing their investment in planting and raising the trees, due to reclamation of the land use by the legal owner before the tree can yield benefits. In some countries, tree planting by tenants is not tolerated by landlords, because it constitutes a symbol of land ownership. Since illegal occupation of land is a frequent practice in many urban areas of the global South, as well as in the global North, the gardens established on such land tend to have short-term investment characteristics, dominated by annual crops (FAO, 1995). Nevertheless, many possibilities exist for arrangements to enable urban dwellers' access and exclusive *usus fructus* rights to garden areas and trees on public and open access land, which is often exposed to destructive harvesting and vandalism.

18.2.2 Urban gardening vs. urban horticulture, agriculture and agroforestry

For inner-urban gardens, various terms are in use, and there are no sharp definitions in place. Urban "gardening" (or *horticulture*, the Latin term for garden) overlaps largely with urban "agriculture" (city farming), especially when it comes to smaller plot sizes. However, it may be possible to differentiate with regard to the tree component. Trees are considered a very integral part of gardens, while at the core of conventional agriculture, (mono-) cropping of non-woody food plants is typical. Nevertheless, the delineation of these terms rests deeply in the cultural and professional ties of the authors.

Several of the practices under focus could also be described as urban "agroforestry systems" (Odurukwe, 2006), highlighting the use of basic agroforestry techniques in urban gardens – for example, growing woody and non-woody crops in spatial arrangement and/or temporal sequence, with ecological and economical interactions between the different components (Nair, 1993). Fruit trees are often cultivated by both urban dwellers in the global South (FAO, 1995) and in the global North. In addition, certain recent developments in urban gardening, such as permaculture or integration of livestock components, comply with agroforestry principles.

A number of denoting features of the phenomenon "urban gardening" are collected from Arndt and Haidle (2004). "Urban gardening" takes place in intra- and peri-urban areas, on land or constructed plains (e.g., roofs), which were not originally designated for gardening. The source of labor, material inputs, as well as the consumption of both the highly diverse edible and non-edible products, is largely local. The quantities produced are small. Besides growing plants and raising animals, the gardens offer an element of space and have positive social impacts. Furthermore, working with plants and the subsistence production of food has psychological, pedagogical and societal meaning.

18.3 Urban gardens

18.3.1 Generic types of urban gardens

Under "urban gardens", various types are subsumed. "Private" gardens can be found either directly attached, or set apart from, a dwelling place, and in a diversity of designs ranging from primary food productions to purely park-like arrangements with only amenity purposes. Non-private gardens incorporate several models:

- *Allotment garden* complexes contain individually cultivated garden plots, with a collectively organized infrastructure and a formal organization. The plot size ranges mostly between 50 and 400 m², and is often fenced and furnished with a summer house or shelter. Gardeners have to follow regulations on utilization, especially regarding the use of a minimum share for subsistence crop production, and the prohibition of use as a permanent residence. Usually, the land is earmarked as non-developable, and is owned by a public or non-profit corporation. Gardeners pay a small fee as rent. This model has been well established in the industrial areas of Europe since the 19th century and, due to its important social functions, national laws have been enacted for the protection and regulation of utilization and subvention.

- *Communal gardens* are park-like green areas, privately or jointly owned and maintained, and usually with regulated access for the public. This model is very common in the United Kingdom.
- *Community gardens* are found on both private and public land and are cultivated by a cooperative group of people, with individual or shared beds for growing food and/or ornamental plants. This kind of garden is found in neighborhoods (*neighborhood gardens*), residential housing grounds (*residential gardens*), adjoined to institutions (*institutional gardens*) such as hospitals, or in educational and recreational settings such as schools (*demonstration gardens*) (University of California, 2014). Most community gardens are traditional public parks open to the public, providing opportunities for recreation, social gatherings, beautification and education. The term is more commonly used in the USA, Canada, Australia and New Zealand (GOAL, undated).

18.3.2 Urban gardens in history

The earliest historical records describe gardens as multi-purpose spaces providing flowers, fruits, shade and food (Bennis, 2006). The first cities in the Middle East, China and Latin America integrated local food production as a resilience strategy for times of conflict (Prain, 2006). Homer, in his epic poems, mentions a paradise-like garden with several kinds of fruit trees (apple, fig, pomegranate, pear, vine, olive). Roman houses were designed to incorporate an inner courtyard as a garden (*peristyle*), providing plants for pleasure, food, dyes, medicines and so on, and surrounded by a covered walkway. The garden was used for privacy, rest and relaxation, while another courtyard (*atrium*) was used for more formal purposes.

In the ancient high cultures of Assyrian, Egyptian, Greek and Roman cities, trees were planted as elements of cultic complexes and temples. Succeeding empires abolished the "pagan" temple complexes, together with their trees, in order to eradicate these religious cults. This public repression of trees was continued in European cities up to the Middle Ages; trees were usually only to be found in private gardens, behind the house, physically and visually separated from openly accessible areas (Lawrence, 2006).

These gardens were regarded as pleasant places (*locus amoenus*), for the exclusive use of the owner, as a refuge in idyllic, cultivated nature amongst plants that provided fragrances, herbs, medicines, fruit, vegetables, flowers and so on. The enclosures were walled to separate them from public areas (*hortus conclusus*). They had high symbolic and cultural significance, such as an association with the Garden of Eden. The trees planted here were protected from the harsh daily life of public roads and places, as well as from damage caused by vehicles, browsing animals, urine and vandalism. The few trees in public open spaces only stood in functional, central locations, such as the *Tilia* trees for jurisdiction sessions. Trees were also found in monastery gardens, cemeteries, vineyards, orchards and, of course, outside the city walls (Lawrence, 2006; Hennebo, 1982). The public green, open spaces which were developed in the same cities, some centuries later, have their roots in these original house gardens.

These origins highlight the significance of a garden, as opposed to wilderness – a sovereign territory of the gardener with planted, designed nature (Rasper, 2012).

18.3.3 Urban gardens for the disadvantaged in the 20th Century

In Germany, in the early 19th Century, "paupers' gardens" (*Armengärten*) emerged in several federal states. For the German city of Kappeln, close to the Danish border, it is reported that the city's commissary for poverty affairs requested the royal land representative to either extend the cramped poorhouse, or to provide land for "mini-farms" (*Kohlhöfe*) to settle the poor. The latter was granted. This model seems to have been successful; some years later, additional royal land was divided into 32 garden parcels, and leased out to "garden-less" citizens.

In 1812, the continuing high level of poverty (Western Europe was suffering under the Napoleonic Wars and dynamic population growth), drove the town's pastor, Schröder, to take up this model and to transfer church land, originally provided for his own subsistence, to the poor. He gave 24 parcels with approximately 1000 m² to very poor families as hereditary leasehold. Two years later, he leased out another 24 parcels, this time with a contract requesting the lessees to elect a representative committee for organizing and supervising the complex, as well as for arranging payments and substitutes for gardeners who lacked liquid assets (KlGV Kappeln, undated). This approach to fight hunger and poverty spread quickly and, in 1826, this kind of gardening could be found in 19 German cities.

Emerging industrialization drew rural people into cities, despite poor housing and living conditions. To improve the ill health of workers' children, Dr. Schreber, an orthopedist, realized Friedrich Fröbel's 'kindergarten' idea by establishing a garden in the fringe of the city of Leipzig, where children can play in the fresh air for multiple benefits in health, learning and child development. The idea was successful, and was soon extended to include education in gardening. Beds were demarcated on the fringes of the playground, and each child was responsible for cultivating a bed. However, the children's initial enthusiasm quickly waned, so their parents took over the beds and developed them with fences and a shelter and, in 1870, the first "Schrebergarten" complex (allotment gardens) was formed in Leipzig (Pehle and Handwerker, 2008). It was especially in order to balance the unhealthy working and living conditions of workers' families that the so-called "worker's gardens" were organized by NGOs (e.g., the Red Cross), large industrial firms, railway companies and so on.

At the turn of the century, the spread of tuberculosis was responsible for increasing the establishment of "gardens for health" in Germany, which rose to over 30,000 in 1911 (DRK, undated). The fast-growing railway network divided the field layout of many peri-urban farms. Isolated parcels of agricultural land were sold off and were made available for gardens.

The origins of "Höhenluft 1", one of over 370 allotment garden complexes in Dresden, show that the gardeners mostly came from the lower social strata. They were suffering from ill health and sought a means for obtaining food, a healthier environment, nature and privacy. The founder was a simple chair maker, who was chronically sick and hence forced to diversify his livelihood. When the expanding city of Dresden swallowed his small garden, he leased a larger field outside the city in 1912, and organized 27 of his peers to form a joint garden complex. The gardeners produced solely staple crops and small livestock for their own use on their plots. The chair maker kept the exclusive right to run a bar for his additional income. A formal association was founded in 1917. When, by 1940, the complex had grown to 4.7 ha, the association purchased all the land in order to protect it from speculation and overriding municipal planning (KGV Höhenluft I, 2013,

Figure 18.1 Allotment garden complex Höhenluft 1 in 1932 (above) and 2002 (below), in Dresden, Germany (KGV Höhenluft 2013. Reproduced with permission of KGV Höhenluft).

pers. comm. Thomak, S. and Völkel, B). The photographs in Figure 18.1 show the development of the central part of the complex, with its remarkable incorporation of trees.

The great wars in the first half of the 20th century pushed all gardeners, on both sides of the conflict, to support national food production. Governments launched campaigns like the German *Winterhilfe* (winter aid), the US *Victory Gardens*, or the UK's *Dig for Victory*. For many, the allotment garden was their only means of survival, and also their only shelter after bombing, displacement or destruction. After World War II, the German

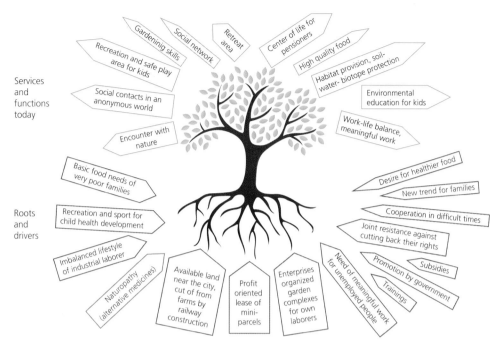

Figure 18.2 Drivers and benefits from allotment gardens over time (Artwork by Naomi Auch. Reproduced with permission).

Democratic Republic nationalized all garden complexes on its territory, and both the establishment of a central organization and the provision of subsidies formalized their strategic role in complementing the national food production. Nationwide, thousands of tonnes of fruit and vegetables, as well as rabbits, eggs and honey, were annually produced by allotment gardeners (Leppert, 2009).

To return to Höhenluft 1, for many gardeners, their parcel became a space where they could enjoy privacy and a network to create celebrations that were otherwise not possible. In recent decades, the use of gardens for recreation in a natural environment, for social encounters and as space for children, has complemented their use for the production of food. The latter has undergone a transition from food as a means to survive, towards achieving tasty and healthy food, as opposed to the conventional "outside-nice" but "inside-poor" supermarket offers. Current topics include the struggle against municipal investment plans, which aim to convert gardens into building sites. Additionally, children's education, nature conservation and the integration of migrants are important issues for the purpose of gardens.

The case of allotment gardens demonstrates the changing drivers and benefits of urban gardens over the period of a century (Figure 18.2). The relevance of gardens as an instrument to solve burning issues of each period is surprisingly high.

18.4 Function and benefits/services of trees and gardens in urban contexts

A terrestrial assessment of the role of trees in Höhenluft 1 garden plots shows that approximately 15% of the survey's garden area is under tree cover (Figure 18.3), composed of both fruit and ornamental trees. The majority of gardeners have a clear idea of

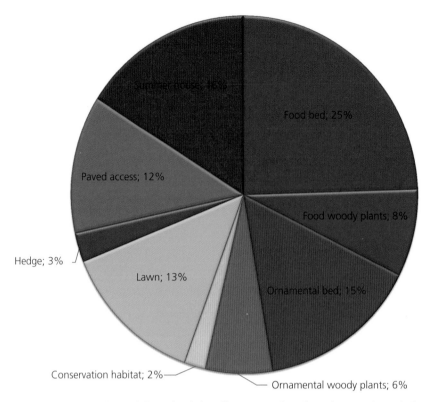

Figure 18.3 Average composition of the Höhenluft 1 allotment garden plots, shares estimated (data source: Survey of 47 garden plots in 2013, average plot area 240 m²).

what proportion of their gardens should be dedicated to trees – neither too small, nor too large. In the case of a conflict, about one-third of the gardeners would be willing to sacrifice crop beds for more trees. The gardeners in the examined group perceive a share of 15% trees as beneficial and comfortable (Fotschki, 2014). This result confirms the findings of McIndoe and McIndoe (2009), that the presence of trees in the world of humans is so self-evident that they are not consciously aware of it.

18.5 Recent forms of urban gardening in the global North and global South

18.5.1 Factors facilitating the emergence

Food growing has become neglected in the global North's cities, due to industrial food production and urban development regulations. Today, gardens are returning to the cities, especially where the peak of industrial growth has passed, and structural change has released both space and the challenge for new ideas. The revitalization of urban gardening is driven by new political convictions, environmental pollution and diminishing fossil fuel resources. It often constitutes resistance against a predominating neo-liberal order, and is a political demonstration for a future worth living in. In the global South, continuing traditional garden practices are complemented by innovative models utilizing the advantages of urban areas for food production. Enabling factors are:

- Post-growth situation. Due to structural change, space (and labor) are more easily available, and scarce municipal financial budgets give room for civic engagement. Conventional fruit and vegetables from agro-industrial mass production are poor in taste and nutritional value, compared with those grown traditionally in home gardens.
- The prevailing economic paradigm expects people to only consume in their free time. Urban gardening overrides this rule, by using free time to grow plants and improve skills while, at the same time, producing supplemental food, which has an emotional significance. Sometimes, this is combined with a spiritual dimension of (joint) experience and practice in nature (Richard, 2012).
- The practice of "mobile gardening", with cultivation in raised beds, re-used juice boxes, soil-filled sacks, tins or other containers, allows growing to be independent of polluted urban soil, and enables the temporary use of space, even where there is a latent risk of being forced to leave the garden in the near future. It demonstrates a nomadic *avant-garde* in economics and arts, well informed by media and highly interconnected via the internet (Werner, 2012).
- Beyond the established institutions and typically represented by "guerrilla gardening", people like to go their own way which, in turn, stimulates institutions into thinking forward. The municipality of Dortmund thus encourages "guerrilla gardeners" to join co-management agreements for public green space (Steffens, 2013).
- Urban gardening in the global North has become an attractive lifestyle. Urban gardens are social institutions, simultaneously public and private, used for relaxation and for actively meeting peers, in the middle of the city. In contrast to their historical predecessors, they are part and parcel of the urban world, providing experiences instead of events in the event-jungle of the city (Borgstedt, 2012). The drivers for the movement are (ibid.):
  1 *Re-grounding*: the "fun and adventure-oriented society" is becoming tired. It longs for stability and predictability, likes to retreat from the complex world with its permanent and compulsory decision-making, and realizes this retreat in the near social environment, traversing different social milieus.
  2 *Autonomy*: people caught between public demands and personal refuge yearn for the feeling of belonging to a larger social group, but with the simultaneous freedom of being able to decide to go their own, new ways.
  3 *Civic engagement*: urban gardeners are responsive, caring for public welfare in their grass root movements, aiming not only at doing things right, but also at doing the right thing.
  4 *Sensuality and diversity*: focus lies on the relevant and the feeling of being self-made, a counterpoint for rationalism and alienation in everyday life. It transforms daily routines into sensual/delightful rituals, enriching daily goods with smell and taste, searching for originality. Urban gardening enables emotional products, produced by oneself.
- Urban home gardens are seen as a strategy of the urban disadvantaged to cope with problems (lack of food, health, urban green), especially in situations where land (space), labor and water is available. On a larger scale (urban agriculture), this can serve as economic revitalization.

18.5.2 Newer urban garden forms (selection)

Guerrilla gardening is seen as the mother of modern gardening movements. The term appeared in the 1970s in New York, with the "Green Guerrillas" – now an established non-government organization or NGO (www.greenguerillas.org). Their activities (Fig. 18.4)

Figure 18.4 Guerrilla gardening as civic engagement, informal but not illegal (Hanoi).

played a crucial role in the politically motivated gardening movement, which gave rise to many community gardens. Guerrilla gardening is often informal, but is not illegal (Von der Halde *et al.*, 2012).

Garden initiatives, such as neighborhood gardens, multi-generation gardens, international gardens, and so on, often constitute "laboratories" for innovative models of a post-growth society, combining the inclusion of all social strata, peace and harmony with urban subsistence. These "modern commons" continue the community garden idea with individual modifications, and often with a political dimension – for example, women's gardens.

"Edible cities" are cities which deliberately plant edible plants to be harvested by citizens, such as olive trees in Adelaide (FAO, 1995) or walnut trees in Ravensburg. Every year, the municipality of Andernach in Germany plants different kinds of vegetables on public greens (Andernach, 2014). Similarly, *urban food forestry* promotes the utilization of public space for growing perennial crops, such as fruit tree orchards (Clark, 2015).

Thematic/demonstration gardens are established with new, upcoming themes, including permaculture, children's farms, (odor) gardens for disabled people, or for home gardening of nutrient-rich food to complement HIV/AIDS therapies.

Allotment gardens can be innovative for urban areas, where the exchange of commodities is limited due to external factors. Armed conflicts around the city of Popayan in Colombia caused both the migration of people from the surrounding villages to the city, as well as a general down-turn of commodity exchange, including food stuffs. Peri-urban dwellers were trained in home food production on public land to fill the food supply gap.

Urban agroforestry is the continued tradition of home gardening, especially in the humid tropics, where enough light and water allows multiple storages of crops. One of the many initiatives is the Vi Agroforestry training centre, established in several cities in Africa, which operates in both rural and peri-urban areas (www.viagroforestry.org).

18.6 Conclusions

Across the millennia, humans have maintained/created patches of managed nature within their urban living space. These urban gardens show high, mostly multi-functional utility, serving actual needs. Urban gardens have been used as a means to cope with the burning issues of each period, and their development does not stop. Urban gardens, in their diverse designs, continue to be highly appreciated elements in the cities of the global North and global South, enabling both benefits from nature and from the urban environment.

Trees constitute an integrated part of urban gardens. Since urban gardening employs many of the practices described under agroforestry, research in this approach might continue to create useful knowledge to adopt urban gardening towards the many future challenges.

References

Andernach (2014). *Die Essbare Stadt. Municipality of Andernach*. www.andernach.de/de/leben_in_andernach/es_startseite.html [accessed Feb 2015].

Arndt, C. and Haidle, I. (2004). Urbane Gärten in Buenos Aires. Thesis (unpublished), Technische Universität Berlin, Berlin. http://anstiftung.de/jdownloads/Forschungsarbeiten%20Urbane%20Grten/vdhaide_urbanegaerten_hi.pdf [accessed Feb 2015].

Bennis, E.M. (2006). Productive gardens. http://cmsen.eghn.org/etpg-introduction [accessed Feb 2015].

Borgstedt, S. (2012). Das Paradies vor der Haustür: Die Ursprünge einer Sehnsucht aus der Perspektive soziokultureller Trendforschung, pp. 118–125. In: Müller, C. (ed). *Urban gardening. Über die Rückkehr der Gärten in die Stadt*. Oekom, München.

Carreiro, M.M. and Zipperer, W.C. (2008). Urban forestry and the eco-city: today and tomorrow. In: Carreiro, M.M., Song, Y.-C. and Wu, J. (eds). *Ecology, planning, and management of urban forests: international perspectives*, pp. 435–456. Springer Science+Business Media, New York.

Clark, K.H. (2015). *Urban food forestry*. http://urbanfoodforestry.org/about/ [accessed Feb 2015].

DRK (undated). *Arbeitergärten als Tuberkulose-Prophylaxe. German Red Cross*. http://150jahre.drk.de/ [accessed Feb 2015].

FAO (1995). *The potential of urban forestry in developing countries: a concept paper*. Food and Agriculture Organization of the United Nations (FAO). www.fao.org/docrep/005/t1680e/t1680e00.htm [accessed Oct 2013].

Fotschki, L. (2014). *Funktionen von Bäumen in Kleingärten* (unpublished). Technische Universität Dresden, Institute of International Forestry and Forest Products, Tharandt.

GOAL (undated). *Glossary*. www.goal-2025.com/glossary/ [accessed Feb 2015].

Hennebo, D. (1982). Städtische Baumpflanzungen in früherer Zeit. In: Meyer, F.H. (ed). *Bäume in der Stadt*, pp. 11–45. Ulmer, Stuttgart.

KlGV Kappeln (undated). *Vereinsgeschichte. Kohlhöfe (1800) – Carls-Gärten (1806) – Kleingärten (1814)*. www.klgv-kappeln.de/verein/history.html [accessed Feb 2015].

KGV Höhenluft I (2013). *Wir über uns*. http://mk-atomy.de/hoehenluft/index.php/verein-2/geschichte/ [accessed Aug 2013].

Lawrence, H.W. (2006). *City trees: a historical geography from the Renaissance through the nineteenth century*. University of Virginia Press, Charlottesville, London.

Leppert, S. (2009). *Paradies mit Laube. Das Buch über Deutschlands Schrebergärten*. Deutsche Verlagsanstalt, München.

McIndoe, A. and McIndoe, R. (2009). *Bäume für jeden Garten: Ein kreativer Ratgeber*. Delius Klasing, Bielefeld.

Nair, P.K.R. (1993). *An introduction to agroforestry*. Kluwer, Dordrecht.

Odurukwe, S.N. (2006). Agroforestry in peri-urban cities of Abia State, Nigeria. In: Veenhuizen, R. van (ed). *Cities farming for the future. Urban agriculture for green and productive cities*, pp. 435–437. International Institute of Rural Reconstruction; International Development Research Centre, Silang, Philippines, Ottawa.

Pehle, T. and Handwerker, M. (2008). *Unser Schrebergarten*. Moewig, Hamburg

Prain, G. (2006). Participatory technology development for sustainable intensification of urban agriculture. In: van Veenhuizen, R. (ed). *Cities farming for the future. Urban agriculture for green and productive cities*, pp. 275–298. International Institute of Rural Reconstruction; International Development Research Centre, Silang, Philippines, Ottawa.

Rasper, M. (2012). *Vom Gärtnern in der Stadt: die neue Landlust zwischen Beton und Asphalt*. Oekom, München.

Richard, U. (2012). Urbane Gärten als Orte spiritueller Erfahrung. In: Müller, C. (ed). *Urban gardening: Über die Rückkehr der Gärten in die Stadt*, pp. 225–234. Oekom, München.

Steffens, H. (2013). *Stadtblümchen: Gebrauchsanweisung für den kleinen Garden-Guerillero*. www.labkultur.tv/video/stadtbluemchen-gebrauchsanweisung-fuer-den-kleinen-garden-guerillero [accessed Dec 2014].

University of California (2014). *Community Gardens*. http://ucanr.edu/sites/MarinMG/Community_Service_Projects/Marin_Community_Gardens/ [accessed Dec 2014].

Von der Halde, E., Halder, S., Jahnke, J. and Mees, C. (2012). Guerilla Gardening und andere politische Gartenbewegungen. Eine globale Perspektive. In: Müller, C. (ed). *Urban gardening: Über die Rückkehr der Gärten in die Stadt*, pp. 266–278. Oekom, München.

Werner, K. (2012). Eigensinnige Beheimatungen. Gemeinschaftsgärten als Orte des Widerstands gegen die neoliberale Ordnung. In: Müller C. (ed.) *Urban gardening: Über die Rückkehr der Gärten in die Stadt*, pp 54–75. Oekom, München.

CHAPTER 19

Urban woods for relaxation and inspiration

Eckhard Auch[1], Hubertus Pohris[1] and Markus Biernath[2]

[1] Technische Universität Dresden, Tharandt, Germany
[2] Staatsbetrieb Sachsenforst, Forest District Dresden, Dresden, Germany

19.1 Introduction

The worldwide increase in urban population, followed by the expansion of urban areas, currently exerts great pressure on existing urban and peri-urban landscape structures. With this in mind, the management and governance of urban landscape, such as public parks, playgrounds, residential green spaces and adjacent forests, will become progressively challenging, and there must be an increased focus on sustainable urban development. Thus, different aspects represented by a set of social, cultural, ecological and economic attitudes need to be considered, resulting in a multidisciplinary approach (Jansson and Lindgren, 2012).

One important factor of urban landscape management is characterized by urban forestry. This chapter is, therefore, devoted to a discussion of how this integrative and innovative approach can be applied to manage urban forest resources, thus providing multiple benefits to urban society. The development and definition of the urban forestry concept has been described by Konijnendijk (2003) and Konijnendijk *et al.* (2006), with the initial approach in North America, and the concept is also gaining a hold in Europe. In conclusion, urban forest resources are of great importance here, in terms of providing goods and services to society, although outdoor recreation opportunities often have priority over them all. Basic considerations and principles essential for the elaboration of the urban forestry concept are illustrated by using as case the functions and services, silvicultural and management specifics of urban and peri-urban forests around the city of Dresden in Germany.

However, international debates currently centre increasingly on the concept of multiple-use forest management for timber, non-timber forest products (NTFPs) and environmental services in the tropics (García-Fernández *et al.*, 2008; Guariguata *et al.*, 2012). It seems it will only be a matter of time before this discourse also broadens to include a discussion of multiple-use forest management for urban and peri-urban landscape structures in the tropics.

19.2 Some definitions

The term *urban* refers to "the entire area in which a city's sphere of influence (social, ecological and economical) comes to bear daily and directly on the population" (Schiere, 2006). It highlights the usual high density of people, buildings and constructed infrastructure,

Urban Tree Management: For the Sustainable Development of Green Cities, First Edition. Edited by Andreas Roloff.
© 2016 John Wiley & Sons, Ltd. Published 2016 by John Wiley & Sons, Ltd.

as well as their manifold interactions. MEA (2005) considers settlements with over 5000 inhabitants as urban. Its antonym is *rural,* characterized by a close relationship with nature, a lower population density and good social networks (Pretzsch, 2014). Peri-urban (or "r-urban", FAO, 2013) areas present a mixed zone on the urban fringe, combining rural with urban elements, but functionally strongly connected to the urban centre.

The terms *forest* and *woods* are used interchangeably, and refer to the same type of managed ecosystem. The Latin roots of the term *forests* (foris) convey a meaning of "reserve with limited/exclusive utilization", whereas the roots of *wood* refer to a "collection of trees" in the landscape (Konijnendijk, 2008). Some schools prefer to use the word *forest* to highlight the production function, and *woods/woodland* to imply naturalness and biodiversity.

Urban forests constitute a mosaic of trees and other vegetation which are affected by urban conditions, and which are managed intensively by different agencies and people (Carreiro and Zipperer, 2008). The term *urban forestry* was coined in the 1960s in a Canadian study on tree and forest management for the benefit of urban society. The concept is characterized by a spatial focus on urban and peri-urban, as well as by an integrative, participatory approach, a strategic and long-term perspective, interdisciplinary concepts of applied, natural and social sciences, and a multi-functional focus (Konijnendijk, 2005). It is a "type of forestry that serves to improve the quality of human life in the city" (Da and Song, 2008).

It is a challenge to provide a clear rendering of the term *urban forestry* when considered in terms of the extensive scope described. Konijnendijk *et al.* (2006) define urban forestry "as the art, science and technology of managing trees and forest resources in and around urban community ecosystems, for the physiological, sociological, economic and aesthetic benefits trees provide society". In the context here, the individual tree managed along streets, in median strips and in parks, results in *arboriculture*, whereas urban, peri-urban or rural forest ecosystem management is implemented by *silvicultural* operations.

In the latter case, the trees form dense stands as a predominant living form. Stand-relevant interactions occur between them, affecting their growth and development behavior, and producing a specific forest microclimate, as well as a characteristic forest soil profile. The distribution, composition and structure of urban forests on the one hand, and their dynamic living processes, with planned interventions or unplanned natural disturbance events, on the other, ultimately create a variable set of multi-purpose functions and benefits. In urban forests, the careful balance of their value for recreation, biodiversity and landscape character is at the fore (Tyrväinen *et al.*, 2005). In peri-urban forests, wood production may also be embodied in a system for silvicultural management of forests within urban landscapes (Von Gadow, 2002).

19.3 Forest ecosystem functions and services

The basic idea behind the concepts of *forest functions* and *ecosystem services* are the (forest) ecosystems which provide functional/beneficial services to the human system. The concept of "forest functions for society", apart from delivering timber and fuel, emerged about a century ago. It was discussed by German-speaking forestry academia, and became popular with the paradigm shift in the 1970s from production as a prime function (which automatically generated the desired ecological and social functions), towards a multi-purpose approach with recognition of the equal importance of multiple functions. The concept of

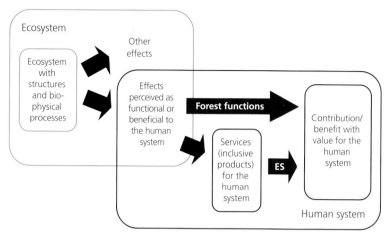

Figure 19.1 Functions and services from the ecosystem for the human system (ES = Ecosystem Services. Under use of AGFE-AGLP, 1982; Haines-Young and Potschin, 2009; De Groot *et al.*, 2010).

"Ecosystem Services" emerged in the USA in the 1970s, and its usage by the Millennium Ecosystem Assessment (2005) has promoted it to an international standard (AGFE-AGLP, 1982; Pistorius *et al.*, 2012).

Lastly, the two concepts use the same idea, and highlight the same causal link, with different technical terms. Both use the ecosystem and the human system and, of all the services generated by an ecosystem, only consider those that are appreciated by humans as being "functional", or are seen as "services" for the human system (Figure 19.1).

An ecosystems' benefits appreciated by humans are products of the manifold biophysical processes within the ecosystems themselves. These different processes are determined by complex physical and chemical interactions between an ecosystems' biotic and abiotic components. They are movements that are building, decomposing and influencing the flows of the system's structures, elements and organisms.

To contrast generally non-marketable services with marketable forest products – typically timber – the term "forest function" was coined. This refers to the impacts of forests, with their contributions to the superior social system (AGFE-AGLP, 1982). Along the same lines, De Groot *et al.* (2002) defines "ecosystem functions" as "the capacity of natural processes and components to provide goods and services that satisfy human needs, directly or indirectly". This concept is used to quantify and to summarize the total value of all ecosystem services in monetary terms (MEA, 2005).

The set of ecosystem services is more universal (see Figure 19.2) than the sets of forest functions. The latter are more operational, and refer to acknowledged, local/regional benefits from forest ecosystems. Forest functions are a tool for designing and guiding multipurpose forest management. They are categorized under productive, protective and socio-economic functions (FAO, 2010), with national specifications (e.g., recreation and welfare in Austria – BMLFUW, 2006). A detailed set of forest functions used in Saxony is given in Table 19.1.

Forest function mapping in Saxony is done on very detailed level (scale 1 : 25 000). Mapping of obvious functions which occur in all forest areas, such as carbon sequestration, timber/wood production, oxygen provision, and so on, are not mapped. The objects

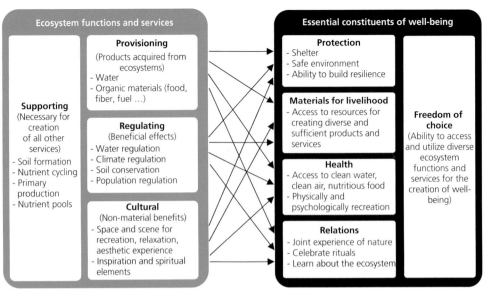

Figure 19.2 Ecosystem services and their links to the human system. (Adopted from MEA 2005).

Table 19.1 Categories of forest functions for Saxony, Germany (selection).

Thematic area	Legally binding		Not legally binding
	Status given to individual areas only after formal declaration	Status given by law	Status designated by forest function mapping
Soil	Infrastructure protection forest, Road protection forest	Soil protection forest	Forest with special protection function for infrastructure/soil
Water	Water protection area, Spa protection zone, Floodplain 1, Water protection forest	Floodplain 2	Forest with special protection function for water/flood protection
Air	Climate protection forest, Emission protection forest, Noise protection forest	–	Forest with special protection function for local/regional climate protection, emission protection, noise protection
Nature	National park, Natura 2000 (Fauna-Flora-Habitat, Bird Special Protection Area), Nature reserve, Natural monument, Natural forest sanctuary, Wildlife reserve	Protected biotope	Forest with special protection function for biotopes, Residual forest in a forest-poor region, Rehabilitation forest, Gene conservation forest, Seed production forest, Forest for education, Forest with special function for fire control
Landscape	Landscape protection area, Protected landscape element	–	Landscape determining forest, Forest with special masking function
Culture	Biosphere reserve, Heritage conservation area, Excavation reserve, Archaeological reserve	Cultural monument	Forest with special heritage conservation function, Documentation area of historical silvicultural practices
Recreation	Recreational forest, Nature park	–	Forest of special recreational function: intensity levels 1 and 2

The forest functions are structured in seven thematic areas (rows), and according to the legal base (columns) for the function (adapted from LFP 2004).

of mapping are the special, exceeding normal level functions (see Table 19.1). All the different functions from an area are documented, and multiple functions are mapped as overlapping.

Forest function mapping is like an inventory, mapping features and functions of high relevance for contemporary society. Conflicting functions are not ranked; it rests with the managers to design balanced compromises with minimized trade-offs. The functions are assessed and mapped for all forests, whether under public or private ownership. Forests close to urban areas usually have to deliver services on a larger scale than those in rural areas, due to the demands of a larger number of users.

In a multi-functional/multi-purpose respect, these forests simultaneously deliver both products as well as protection and recreation services (functions). Forest function mapping supports forest managers in tailoring management interventions that facilitate the natural processes, delivering the requested and appreciated services for society (LFP, 2004; Sachsen.de, 2013).

19.4 Changing demands on urban and peri-urban forests – the case of Dresden

19.4.1 Change in forest functions

Humans have been shaping central Europe's forests for thousands of years. Over the course of time, societal demands towards these forests have changed, causing the various forms of forest utilization to continually adapt. Some of the changing functions of forests for German society are described by Hasel (1985). Accordingly, in the early Middle Ages, the large forest areas were perceived inconsistently as cultural barriers, and also as reserves of land for political and economic purposes, such as royal hunting or clearing for settlements. With shrinking forest resources and increasing demands for pasture, bedding, resin, ash, honey, charcoal, fuel and timber, the primary function changed towards the supply of goods. Timber formed increasingly the base of existence for both the subsistence of the people and for their commercial activities.

With industrialization, migration and population growth, as well as the urbanization and pollution of the environment, the functions of forests for the maintenance of landscape and public health came to the fore. Nowadays, a whole bundle of forest functions have to be considered in forest management. Changing political and economic frame conditions, as well as societal demands, are going to push the shift of forest functions further, across all categories of forest ownerships.

19.4.2 Functional transformation of the *Dresdner Heide* forest, with focus on recreation

The city of Dresden, capital of the German state of Saxony, which was first documented in 1206, became a permanent royal seat in 1485. Since that day, the presence of the royal land has shaped the development of the city and surrounding countryside (Figure 19.3). For centuries, the 50 km^2 forest *Dresdner Heide* (Dresden Heath) was a royal hunting reserve, protected by sovereign forest regulations, and it was in this way conserved with only marginal losses, despite the continuously spreading city of Dresden.

The *Dresdner Heide* has a dense, historically cultivated forest road system. Today, the high recreational value of the *Dresdner Heide* is attributed to its closeness to the city and

Figure 19.3 The Langebrücker Saugarten – a historical hunting enclosure from the 18th Century in the *Dresdner Heide* (Photograph by M. Biernath).

extensive opening up with paths and roads. The layout of the present road network is made up of many irregular crossing paths, with "folksy" names, and identified by historic way-markers on bordering trees. Initially, these paths connected the villages around the forest.

In the 16th century, the road system was extended by roads set in an octagonal diamond layout, with concentric circles. The diamond served for the organization of royal hunting sessions. In the centre, an installation was constructed to catch wild boars (Fig. 19.3).

In 1832, with the introduction of scientific forestry and forest management planning, Heinrich Cotta set up a system of right-angled roads to delimit equally sized compartments. This structure is still valid and is used for forest management today.

The present forest road net results from all three periods. Fundamental changes occurred in the 19th century; the forest became state property, but was still used by the royal family for hunting, and it was increasingly utilized by the inhabitants of Dresden for recreation purposes. This type of recreation in the urban outskirts was boosted in the second half of the 19th century by the construction of tramways and railways to the suburbs.

Booming industrialization attracted immigration. The number of Dresden's inhabitants increased from 100 000 in 1850, up to half a million in 1910. Burdensome work, and crowded housing conditions, plus heavy air pollution, created a strong desire to be in the forest and in the midst of nature on Sundays and public holidays. Forest paths were signposted, and were upgraded to hiking trails and bridle paths. Recreational infrastructure,

Figure 19.4 Weisser Hirsch playground in the *Dresdner Heide* – intensively used by young families at every season (Photograph by M. Biernath).

such as picnic areas and "forest restaurants", were established. Villa suburbs developed around the *Dresdner Heide*, and a sanatorium was built. Hiking and beautification associations were founded to develop the forest for recreation and relaxation activities, thus complementing the state forestry administration's measures.

With the involvement of civil society and private sponsors, about ten forest parks were established. As it was typical for the time, these park-like areas were intensively developed and well maintained, and acted as carefully designed transitions between the city and the true forest. Forest visitors were concentrated in the parks, and timber production became subordinate to the recreational function. The parks were rented to local beautification associations, which acted as sponsors and provided maintenance. They were used for cultural and sporting events and offered an infrastructure, such as places for concerts, covered promenades, tennis courts or sled runs. After the 1920s, the parks were given up, due to societal change and wars. Nowadays, the state forest service is successfully revitalizing some of these parks.

Today, the *Dresdner Heide* is used extensively for recreation. User groups include walkers, hikers, dog handlers, mushroom collectors, joggers, Nordic-walkers, cyclists, mountain bikers, horse riders and geocachers. The forest is visited for recreational activities throughout the day, on a daily basis, all year round (Figure 19.4).

On peak days, several tens of thousands of people pay a visit, and therefore three-quarters of the *Dresdner Heide* is mapped as "high-intensity recreation forest".

19.5 Urban forestry and silviculture

19.5.1 Urban forests as recreational resource

It has recently become common practice in numerous studies throughout Europe to emphasize forests as the most popular setting for outdoor recreation (Nielsen *et al.*, 2007). In fact, their contribution to a high quality of physical and psychological recreation of urban residents is outstanding (Table 19.2).

When dealing with the design aspects of urban forest management, the aesthetics in the range of wilderness (free-growing, non-managed zone) up to even-aged stands (artificially established, intensively managed zone), arises as matter for discussion.

The research results and literature output from North America and Europe on public forest preferences should be interpreted with certain caution, due to varying forest ecosystems, the extensive scale of forest management practice, and also the diverse cultural and social contexts. The public, however, have tended to give high ranking to an uneven-aged, multiple-layered mixture of trees and species, in a stand with different sizes (Stölb, 2005; Gundersen and Frivold, 2008; Qui *et al.*, 2013). All in all, this preference often coincides with close-to-nature forestry. This is partially rooted in the positive cognitive value ascribed to this line of silviculture, because varied forest appearances and naturalness result in high recreational values and an ethical appeal (Nielsen *et al.*, 2007). Moreover, close-to-nature forests commonly provide a remarkable richness of refuges and reproduction habitats for wild plants and animals and, therefore, substantially contribute to biodiversity conservation and biological integration processes (Kimmins, 1997).

19.5.2 Silvicultural operations for recreational resources

The outdoor recreation demands of residents in urban and peri-urban forest areas have to be determined by means of qualified criteria and indicators (e.g., urban population, urban forest location, urban forest visitor count, tourism centers, etc.). The assessment of the opportunities for outdoor recreation services on the basis of the forest's resources (using formative factors and silvicultural options) ultimately leads to the projection of recreational forests, with or without any use for wood production (Table 19.3). To this end, a specific set of silvicultural operations at various single tree, forest stand and forest region levels is identified. It must be directed towards achieving specified goals, objectives and targets, defined by a participatory planning procedure in multi-functional forests.

Table 19.2 Overview of positive forest benefits on recreation and health of the urban residents.

Social potential	
Socio-hygienic (Health and recovery)	**Socio-psychological** (Aesthetics and emotion)
Physical recreation	**Psychological recreation**
• Equable air temperature • Higher relative air humidity • Reduced wind speed • Purified air conditions • Decreased radiation intensity • Less urban noise	• Nature observation • Forest experience • Stress breakdown • Quicker recovery • Joy of life • Thirst for action

Table 19.3 Formative factors and related silvicultural options for the arrangement of recreation forests.

Formative factors	Silvicultural options
Natural conditions	
• Climate	→ Microclimate amelioration
• Geomorphology	→ Windbreak/erosion control
• Forest distribution	→ Landscape projection
• Forest edges	→ Outwards/inwards forest edge shaping
• Forest structure	→ Multiple forest species composition/structures
• Natural curiosities	→ Over-mature tree maintenance/observation spot establishment
Recreational equipment	
• Parking places	→ Intensive zone management
• Food path network • Cycle path network • Bridle path network • Catering, other services	→ Forest visitor guidance/forest infrastructure establishment
• Forest educational path network	→ Forest visitor educational concept
• Seat opportunities • Weather shelter • Observation tower	→ Forest visitor equipment concept
Disturbance regime	
• Air hygienic situation	→ Forest pollution filter management
• Noise hygienic situation	→ Forest marginal zone management
• Forest aesthetic situation	→ Forest protection management

GIS-based planning tools are recommended to facilitate the integration of the recreational functions of forests in the operational forest management planning (Vries and Goossen, 2002).

In general, urban silvicultural activities are oriented towards the treatment of forests within the framework of local site and stand conditions, well-suited to meeting the set of goals prescribed by the public in a sustainable manner, with a minimum effort. Corresponding to this, the principle of deliberate use of the "gratis power" of natural processes must be favored. This line of action is strongly connected to the silvicultural *single tree selection system* or *small group selection system*, which lead to permanent forest stocking. Before elaborating on silvicultural management systems, the planning procedure for each spatially defined land parcel must guarantee an assessment of the given potential, the actual situation, the functions and the objectives (Thomasius, 1992):

• Assessment of the potential is directed towards the site conditions and the natural forest community.
• Assessment of the actual situation is aimed at determining present forest stocking, and further important factors for the intended forest treatment.
• Assessment of the functions follows public requirements, according to participatory planning.

- Formulation of the objectives usually considers close-to-nature forestry with defined tree species composition, and the forest structure in terms of space and age.

The most common operations to be included in the elaboration of the silvicultural management system in the forest establishment phase are: natural and artificial regeneration, tending and weed control, and pest and disease management; while thinning, pruning, and tree harvesting during the forest development phase. These decisions have to be embedded in the urban forest management organization, which ultimately considers the long-term (strategic), medium-term (tactical) and short-term (operational) level (Jansson and Lindgren, 2012).

19.6 Silvicultural specifics of urban and peri-urban forest management

The management and silviculture of peri-urban forests depends on natural and anthropogenious variables. Principles for daily forest management decisions are in place, and are relevant and validated in many urban forests (Weber, 2004). They may not, however, necessarily be applicable in every case (Röhrig, 1980). Lessons learned in the course of research (e.g., Zundel and Völksen, 2002), and the authors' own practical experience, are used to draw conclusions for silviculture and for the organization of urban forest management, in order to address the recreational function and value of urban forests.

Essential frame conditions for the management of recreation forests are made up by:
1 Distance, position and size of the forest with regard to the city.
2 Natural site conditions, tree species composition and forest structure.
3 Type and intensity of recreation activities; and
4 Management and business objectives of the forest owner.

These factors, and their interactions, need to be considered when forming management decisions with deliberate facilitation for visitors' recreation activities. Often, the optimal compromises have considerable trade-offs between the conflicting objectives of stakeholder groups and the available resources of the forest enterprise.

The smaller the area of a peri-urban forest, the more important is its recreational function (Röhrig, 1980). The highest concentrations of visitors are found in small forests within, or directly bordering, the fringes of a city. In these forests, there are usually permanent restrictions in conducting necessary forest management operations due to recreation facilities, such as buildings in the forest, or forest small walking and cycle paths only. Whether or not the forest owner is prepared to accepts this, recreation overrides all other functions and objectives. In more distant and larger forests, the owner may realize a multifunctional concept by harmonizing recreation for visitors with forest management operations. Here, conflicts are infrequent and only temporary, such as road damage after skidding, or exclusion of walkers from a declared mountain bike trail.

The composition of tree species results from both:
1 natural site conditions (which determine the range of possible, adapted species); and
2 earlier utilization at a time when forests were markedly managed towards specific goals.

In peri-urban forests, timber production was often practiced in even-aged monocultures, resulting in homogenous forests without special aesthetic stimulus. Successful examples for transforming forests with silvicultural operations into diverse, attractive recreation forests are the forest parks in the 19th century, or numerous municipal forests such as the one in Lübeck, or the Sihlwald of Zürich.

Opinions about the features that make a forest attractive for recreation are quite diverse and contradictory. Zundel and Völksen (2002) found that, in Germany, preference was given as follows to mixed forest (67%), conifer forest (25%) and purely deciduous forest (10%), but there were also many indecisive responses. The preferences of visitors are subjective, emerging individually from personal experiences as child, the types of forests a person is used to, values and guiding images acquired by enculturation, and so on. Attributes for attractive, aesthetic forests (in central Europe) are structures in both species (mixed forest) and size (age/diameter). Usually, close-to-nature forests serve these requirements best; they deliver positive impressions and facilitate recreation and relaxation. Additional forest-aesthetic aspects are identified by Burschel and Huss (1987):

- Preservation of old and "bizarre" individual trees or groups of trees.
- Planting of rare or charming tree species on exposed areas such as roads, paths intersections or viewpoints.
- Maintenance, and creation or restoration of open spaces such as meadows, ponds, historic aisles or views.
- Intensified silvicultural treatment of stands around rocks, springs, cultural monuments or landmarks, and so on, to emphasize the effect.
- Maintaining or restoring historical forms of forest use (e.g., coppice system or coppice with standard system).
- Preservation and restoration of testimonies of past eras (e.g., alleys, ditches or hunting facilities).

The strongest objections raised by German forest visitors relate to harvesting operations with heavy forestry machinery, resulting in intensively thinned stands and damaged roads. These disturbances are perceived as restricting enjoyment of the forest. Moreover, many visitors are uninformed about forest ecosystem dynamics, and assume that harvesting destroys the ecosystem. It is, therefore, extremely important to inform people about these measures by means of media and public relations work and, of course, to choose practices which are compatible with recreational activities. Some positive examples with reference to the *Dresdner Heide* are:

- Execution of manual logging operations in recreational core zones or close to buildings, using well-trained, experienced forest workers only.
- Skidding of logs in these areas with draft horses (Figure 19.5); horses and demonstrations of traditional techniques are emotional highlights, and are positively perceived by visitors, as well as having a reduced impact on soil.
- Practice of reduced impact logging methods, employment of modern forestry equipment only, extended skid trail distances (40 m), awaiting ideal conditions for skidding (frost or drought) to reduce soil compaction, even at the expense of higher harvesting costs.
- Repair and maintenance of damaged forest roads immediately after completion of logging and timber transport.

A minimum dense network of forest roads and paths is a prerequisite for the recreational function of peri-urban forests, and it constitutes the true basic service provided by forest owners for visitors. The quality of the roads (accessible all year round, for relaxed cycling), eventful routing, and a reliable guidance system with overview boards, road signs and marked trails, are all important. In the *Dresdner Heide*, fine gravel is used as a top layer for forest roads, implicating higher maintenance costs but satisfying the wishes of the many cyclists.

In situations where user groups intermingle, and there is a possible risk of accidents and conflicts (e.g., walkers with horse riders or with mountain bikers), a spatial separation

Figure 19.5 Skidding of logs with draft horse in winter (Photograph by Jöran Zocher – Sachsenforst. Reproduced with permission).

of user groups is recommended. In Saxony, these conflicts are defused by a regulation that allows horse riding in forests only on roads specifically declared for this purpose. The horse also requires a license issued by the forestry office. As regards mountain biking, it is proposed (with the participation of organized mountain biker groups) to establish mountain bike trails. This will limit the risks to other forest visitors, as well as reducing informal, destructive downhill rides.

Furnishing recreation forests with sport, picnic and educational infrastructure was highly fashionable in Germany in the 1970s. However, many of the lessons learned were negative – such as increases in litter, vandalism, safety liability risks, and lack of resources for maintenance. This led to the trend being reversed, and equipment was reduced to the bare necessities.

Infrastructure for recreation can increase the attractiveness of forests, but it is only useful if financial and human resources for maintenance are sustainably available, because quickly the forest service will be blamed for untidy or damaged facilities. Where these resources cannot be guaranteed, the risk of negative reputation must be accepted. Finally, accident hazards due to broken facilities present a considerable risk of safety liability for the forest owner.

While courts in Germany judge the typical hazards in forests (e.g., from falling branches) as being at the user's own risk, accidents related to recreation facilities (e.g., constructed huts, playgrounds or benches) are the liability of the forest owner. Therefore all trees within a radius of one tree-length around these installations have to undergo regular safety assessment. If third parties, such as hiking associations or nature education organizations, ask to establish recreation facilities, they should assume all safety obligations and liability (AID, 2011).

19.7 Conclusions

The process of urbanization, with its increasing and changing social demands and its expansion of urban landscape, has been a worldwide phenomenon since the 20th century. There is no doubt that urbanization will continue to cause dramatic changes in the relationship between human society and the natural environment. Forests in regions with high visitor pressures require a specific forest management concept, which is defined as *urban forestry*.

The multi-disciplinary activities of urban forestry encompass design, participatory planning, sound silvicultural principles and accepted practices, as well as effective monitoring and evaluation of management activities. They are directed towards trees and forest resources in and around urban community ecosystems.

The following are of utmost importance to urban forest ecosystems:

- Assessment according to their potential for site conditions and forest community development.
- Assessment in view of the functions, goods and services expected by the urban residents.
- Sustainable management for forest ecosystem productivity and stability, considering the principal goals, objectives and targets determined by urban participatory planning and group decision making.
- Urban forests in Europe are predominantly characterized by multi-functionality, with focal points of outdoor recreation, water protection and biodiversity conservation. To optimize their services, they must be integrated into planning. Considering other land-use/cover types, forest functions may be used to operationalize planning.
- Long-term strategic planning for spatial and structural design of urban forest, to adapt stand structure to climate change effects and build up spatial layout (corridors, networks, priority areas for forests, etc.) for urban climate, water, biodiversity conservation and recreational services.

However, as urban forests in other regions of the world all have their own specific set of demands which have to be met, urban forest management concepts may differ tremendously. Nevertheless, the general reflections presented in this chapter may be used to illustrate urban forest management concepts within a changing framework.

References

AGFE-AGLP (1982). *Leitfaden zur Kartierung der Schutz- und Erholungsfunktionen des Waldes (Waldfunktionenkartierung)*, 2nd edition. Sauerländer Verlag, Frankfurt a.M.

AID (2011). *Verkehrssicherungspflicht der Waldbesitzer*. Infodienst Ernährung, Landwirtschaft, Verbraucherschutz e. V. (AID), Bonn.

BMLFUW (2006). *Forest Development Plan (WEP): Guideline on content and design – version 2006*. Federal Ministry of Agriculture, Forestry, Environment and Water Management (BMLFUW), Vienna.

Burschel, P. and Huss, J. (1987). *Grundriß des Waldbaus. Ein Leitfaden für Studium und Praxis*. Paul Parey, Hamburg, Berlin.

Carreiro, M.M. and Zipperer, W.C. (2008). Urban forestry and the eco-city: today and tomorrow. In: Carreiro, M.M., Song, Y.-C. and Wu, J. (eds.) *Ecology, planning, and management of urban forests: international perspectives*, pp. 435–456. Springer Science + Business Media, New York.

Da, L-J. and Song, Y-C. (2008). The construction of near-natural forests in the urban areas of Shanghai. In: Carreiro, M.M., Song, Y.-C. and Wu, J. (eds.) *Ecology, planning, and management of urban forests: international perspectives*, pp. 420–432. Springer Science + Business Media, New York.

De Groot, R.S., Wilson, M.A. and Boumans, R.M.J. (2002). A typology for the classification, description and valuation of ecosystem functions, goods and services. *Ecological Economics* **41**, 393–408.

De Groot, R.S, Alkemade, R., Braat, L., Hein, L. and Willemen, L. (2010). Challenges in integrating the concept of ecosystem services and values in landscape planning, management and decision making. *Ecological Complexity* **7**, 260–272.

FAO (2010): *Global forest resources assessment 2010. Main report.* FAO Forestry paper 163. Food and Agriculture Organization of the United Nations (FAO), Rome

FAO (2013). *Towards the assessment of trees outside forests.* Food and Agriculture Organization of the United Nations (FAO). www.fao.org/publications/card/en/c/a80d6068-ea82-5a47-af20-4ec983f90276/ [accessed March 2015]

García-Fernández, C., Ruiz-Pérez, M. and Wunder, S. (2008). Is multiple-use forest management widely implementable in the tropics? *Forest Ecology and Management* **256**, 1468–1476.

Guariguata, M.R., Sist, P.L. and Nasi, R. (2012). Multiple use management of tropical production forest: how can we move from concept to reality? *Forest Ecology and Management* **263**, 170–174.

Gundersen, V.S. and Frivold, L.H. (2008). Public preferences for forest structures: a review of quantitive surveys from Finland, Norway and Sweden. *Urban Forestry & Urban Greening* **7**, 241–258.

Haines-Young, R. and Potschin, M. (2009). The links between biodiversity, ecosystem services and human well-being. In: Raffaelli, D. and C. Frid (eds). *Ecosystem Ecology: a new synthesis.* BES Ecological Reviews Series, CUP, Cambridge.

Hasel K. (1985). *Forstgeschichte: Ein Grundriß für Studium und Praxis.* Paul Parey, Hamburg, Berlin.

Jansson, M. and Lindgren, T. (2012). A review of the concept "management" in relation to urban landscapes and green spaces: toward a holistic understanding. *Urban Forestry & Urban Greening* **11**, 139–145.

Kimmins, J.P. (1997). *Forest ecology: a foundation for sustainable management.* 2nd edition. Prentice-Hall, New Jersey.

Konijnendijk, C.C. (2003). A decade of urban forestry in Europe. *Forest Policy and Economics* **5**, 173–186.

Konijnendijk, C.C. (2005). New perspectives for urban forests: introducing the wild woodland. In: Kowarik, I. and Körner, S. (eds). *Wild urban woodlands. New perspectives for urban forestry*, pp. 33–45. Springer Berlin/Heidelberg, New York.

Konijnendijk, C.C. (2008). *The forest and the city. The cultural landscape of urban woodland.* Springer Science + Business Media, Dordrecht.

Konijnendijk, C.C., Ricard, R.M., Kenney, A. and Randrup, T B. (2006). Defining urban forestry – a comparative perspective of North America and Europe. *Urban Forestry & Urban Greening* **4**, 93–103.

LFP (2004). *Waldfunktionenkartierung. Grundsätze und Verfahren zur Erfassung der besonderen Schutz- und Erholungsfunktionen des Waldes im Freistaat Sachsen.* Freistaat Sachsen Landesforstpräsidium (LFP), Graupa OT Pirna.

MEA (2005). *Ecosystems and human well-being: a framework for assessment.* Millennium Ecosystem Assessment (MEA). Washington. www.millenniumassessment.org/en/Framework.html [accessed March 2015]

Nielsen, A.B., Olsen, S.B. and Lundhede, T. (2007). An economic valuation of the recreational benefits associated with nature-based forest management practices. *Landscape and Urban Planning* **80**, 63–71.

Pistorius, T., Schaich, H., Winkel, G., Plieninger, T., Bieling, C., Konold, W. and Volz, K-R. (2012). Lessons for REDDplus: a comparative analysis of the German discourse on forest functions and the global ecosystem services debate. *Emerging Economic Mechanisms for Global Forest Governance* **18**, 4–12.

Pretzsch, J. (2014). Paradigms of tropical forestry in rural development. In: Pretzsch, J., Darr, D., Uibrig, H. and Auch, E. (eds). *Forests and rural development*, pp. 7–49. Springer Berlin/Heidelberg.

Qiu, L., Lindberg, S. and Nielsen A.B. (2013). Is biodiversity attractive? – On-site perception of recreational and biodiversity values in urban green space. *Landscape and Urban Planning* **119**, 136–146.

Röhrig, E. (1980). *Waldbau auf ökologischer Grundlage. Vol. 1: Der Wald als Vegetationstyp und seine Bedeutung für den Menschen.* Paul Parey, Hamburg, Berlin.

Sachsen.de (2013). *Waldfunktionen. Staatsbetrieb Sachsenforst, Graupa OT Pirna.* www.forsten.sachsen.de/wald/132.htm [accessed March 2015].

Schiere, H., Thys, E., Matthys, F., Rischkowsky, B. and Schiere, J. (2006). Livestock keeping in urbanised areas, does history repeat itself? In: Veenhuizen, R. van (ed.) *Cities farming for the futur*, pp. 349–366e. International Institute of Rural Reconstruction; International Development Research Centre, Silang, Philippines, Ottawa.

Stölb, W. (2005). *Waldästhetik, über Forstwirtschaft, Naturschutz und die Menschenseele*. Kessel, Remagen-Oberwinter.

Thomasius, H. (1992). Prinzipien eines ökologisch orientierten Waldbaus. *Der Dauerwald, Zeitschrift für naturgemäße Waldwirtschaft* **7**, 2–21.

Tyrväinen, L., Pauleit, S., Seeland, K. and Vries, S. de (2005). Benefits and uses of urban forests and trees. In: Konijnendijk, C.C., Nilsson, K., Randrup, T. B. and Schipperijn, J. (eds). *Urban Forests and Trees*, pp. 81–114. Springer, Berlin, Heidelberg, New York.

Von Gadow, K. (2002). Adapting silvicultural management systems to urban forests. *Urban Forestry & Urban Greening* **1**, 107–113.

Vries S. de and Goossen, M. (2002). Modelling recreational visits to forests and nature areas. *Urban Forestry & Urban Greening* **1**, 5–14.

Weber, N. (2004): *Walderholung und Erholungswald im stadtnahen Raum. Nutzungsansprüche – Konflikte – Lösungsansätze* (unpublished). Freistaat Sachsen Landesforstpräsidium (LFP), Graupa OT Pirna.

Zundel, R. and Völksen G. (2002). *Ergebnisse der Walderholungsforschung: eine vergleichende Darstellung deutschsprachiger Untersuchungen*. Kessel, Remagen-Oberwinter.

CHAPTER 20

Acceptance for urban trees: Environmental education programs

Ulrich Pietzarka

Technische Universität Dresden, Tharandt, Germany

20.1 Introduction

The development and management of urban green, especially trees, requires a broad acceptance at all levels of local societies, up to the willingness of private landowners to plant trees on their property (Dilley and Wolf, 2013; Jack-Scott *et al.*, 2013). It needs space that is scarce in urban environments, which also implies a considerable investment. Later on, it demands well-planned management regarding all the different interests and, therefore, appropriate resources once again.

On the one hand, broad acceptance and the commitment to urban green in public is justification for political and economical decisions of the authorities in favor of urban trees. On the other hand it leads to awareness and protection of urban trees in urban planning or development, it promotes understanding for hazard management, it helps preventing vandalism and not least it tries to involve local communities in their management (Mincey and Vogt, 2014; and see Chapter 17).

This makes it obvious that raising awareness for the benefits, but also the demands, of urban trees requires important services for their establishment, management and development. In addition, acceptance for forestry management may also be gained and, in any case, environmental education contributes to the development of ecological awareness (Konijnendijk, 2008).

For practitioners in urban tree management, a new and sometimes difficult scope of duties thus develops – education in the broadest sense and, especially, environmental education. Too often, persons who are not educationally trained are confronted, more or less voluntarily, with this new challenge. However, it is obvious that, even with great commitment, this combination may lead to mistakes and failure. Because of this, the final chapter in this book offers some notes on basic requirements for environmental education programs that focus on urban trees and forests.

20.2 Education for sustainable development

The various chapters within this book make it obvious how many, in part contrary, requirements need to be met considering urban trees. First of all, there are the biological-ecological requirements of the trees themselves, that are species- or even cultivar-specific. They need to be harmonized with the interests and demands of people living in

Urban Tree Management: For the Sustainable Development of Green Cities, First Edition. Edited by Andreas Roloff.
© 2016 John Wiley & Sons, Ltd. Published 2016 by John Wiley & Sons, Ltd.

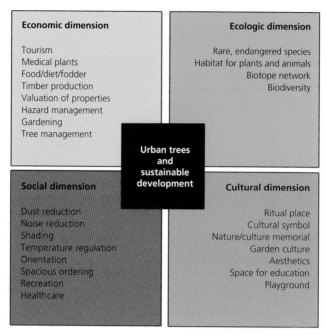

Figure 20.1 Urban tree functions and dimensions of sustainable development (according to Stoltenberg, 2009).

the cities. Figure 20.1 gives a synopsis of the multiple functions and effects of urban trees (Stoltenberg (2009); and see Chapter 1) and their relationship to the dimensions of sustainable development.

All demands for various functions (noise and dust reduction, shading, etc.), protection of monuments, urban planning, traffic control, security, nature and species conservation, just to list a few of the main topics (Chapter 1), are eligible. This is a very typical example of the diversity of interests and demands of society that need to be balanced where the topic of urban trees is concerned (Dilley and Wolf, 2013), and this is exactly what the concept of education for sustainable development is targeted on. It seeks to impart those skills, competencies and knowledge that enable people to actively shape sustainable development.

Therefore, joined-up thinking in the fields of economy, ecology, politics and culture, as well as communicative, social and systematic competencies are necessary (Kohler and Lude, 2010). The so-called "realization competence" which needs to be attained today encompasses twelve sub-competencies (de Haan, 2010). These empower people to identify environmental problems, to evaluate them, balancing different interests, and finally to develop courses of action for their solution:

1 Gather knowledge in a spirit of openness to the world, integrating new perspectives.
2 Think and act in a forward-looking manner.
3 Acquire knowledge and acting in an inter-disciplinary manner.
4 Deal with incomplete and overly complex information.
5 Co-operate in decision-making processes.
6 Cope with the individual dilemmatic situation of decision-making.
7 Participate in collective decision-making processes.

8 Motivate oneself, as well as others, to become active.
9 Reflect upon one's own principles and those of others.
10 Refer to the idea of equity in decision-making and planning actions.
11 Plan and act autonomously.
12 Show empathy for and solidarity with the disadvantaged.

20.3 Features of successful education programs

Are education programs successful only if attendees are able to correct answer questions afterwards? No. This question is easy and clear to answer regarding the previously described concept of education for sustainable development. To impart knowledge is just one (although not unimportant) part of education for sustainable development. It is necessary to stimulate all the competencies listed above. Of course, this is not possible in every single education program, but it becomes clear that it is advisable, and often quite easy, to adapt existing programs to this education concept. In many cases, it is already enough to relate the educational contents to sustainable development and its dimensions (UNESCO, 2012).

To develop and strengthen these competences, education programs and opportunities need to provide special features described afterwards. Some examples of what is needed will help practitioners to implement them.

20.3.1 Specific to target groups

It does not matter who joins an education program – the program, its contents and also its course need to be individually adapted to the knowledge, abilities and interests of the attendees (UNESCO, 2005, 2012). Therefore, it is recommended to gather as much information on the group as possible already before the start of the program. An application form, even if it is filled in by the guides themselves or whoever speaks to applicants on the phone, helps in not forgetting essential questions, such as: theme; number of persons; age; date; time; contact person; address; phone; email; meeting point; special interests. Even, if later on, something goes wrong (and sometimes, somehow, something *will* go wrong), the filled form is useful to prove that one tried one's best to prevent this happening.

Basically, target groups are very individual. However, when preparing an education program, it is advisable to consider a rough classification that reflects the characteristics and abilities of the guests. Such a classification often results from age, or stages of development:

Preschoolers and primary-school pupils are quite easy to inspire. They develop mobility and can let their imaginations run wild. On the other hand, they have lower powers of concentration and limited abilities to read and understand text. Consequently, this has to be considered when designing education programs by alternating play and input or action and concentration. Such a program requires frequent changes of location, and a single stop should not take more than 15 minutes. For this age group, special attention has to be drawn to maximum road safety. Areas with less traffic volume should be visited – for example, public parks.

Adolescents reaching puberty often rebel. They try to find out whether they can exceed limits as they search for their own position in society. They question everybody and everything, but they are also open-minded for something new. This can be a great opportunity. It is crucial to meet them with special respect, and they need to feel they are being taken seriously. The program must not appear as childish play, but as a challenge to be mastered. For at least parts of the program, contests between smaller groups are recommended. Group formation should be randomized (e.g., drawing lots or counting), so that exclusion of single persons is prevented, and an unusual degree of cooperation is needed. This automatically stimulates the development of several sub-competencies (e.g., 5, 7, 10).

For this target group, anything that might be interpreted in a sexual sense must be omitted.

Adults are often stressed out a bit by their jobs and families. They will join education programs only if the contents are personally important, or are directly connected to their job. In the same time, they expect the course to be in a way pleasurable, and also a little entertaining (Ham, 1992), without too much effort. Very often, this target group is reached by offering something for the whole family. Parents enjoy participating in programs more appropriate for children, together with their children, helping and assisting them. The point is to learn together from each other.

Senior citizens usually have more time available. Their children have already left home, and they have perhaps retired. Thus, they have time to spend with new topics, and also in search of further social contacts. This is especially true for intense, urbanized societies, where family ties or social contacts decrease. Senior citizens also tend to share their rich experiences and knowledge with other people of their own, or following, generations. Time for communication is necessary in such an education program. The guests are invited to contribute their point of view (Figure 20.2). This might produce conflicts.

Figure 20.2 Environmental education: lifelong learning for all target groups.

On the one hand, it is good to utilize traditional knowledge and skills but, if general conditions have changed, this knowledge needs to be balanced with new findings. Not everything that was well-tried yesterday is appropriate today, but not everything up-to-date is really new.

Senior citizens might also be a little physically limited. Programs must provide enough time for conversation, and also for short breaks with the possibility to use a rest room. Guests need to know this in order to be prepared and relaxed. If they can be sure that their needs are respected, they will be open-minded for the topic.

20.3.2 Inviting

Attendees of environmental education programs should in all cases be welcomed as dear guests. To commemorate this in the following, they will always be called guests. Guests receive treatment that is friendly, thoughtful and obliging. They enjoy the right to hospitality, and their needs are close to our hearts.

A substantial key to the success of an educational program is the personality of the host, and his/her charisma and appearance. Hosts have to be kind, gentle and polite, but also very well-educated and well-prepared. In some cases they may also assume special roles – what might be more entertaining (see below)? However, if they do not appear authentic, they will not be taken seriously.

This also means that only such persons should perform as hosts who personally identify with the contents and targets of the program, and the institution. They need to enjoy communication with other people – their guests – and do this by choice. Decreed and enforced educational work is not helpful. In too many hierarchical organized administrations, education has been imposed to untrained employees and has ultimately gone wrong.

20.3.3 Focused

The topic "urban trees" risks overburdening and disconcerting guests by trying to impart too much, and too detailed, unsorted information. This can particularly jeopardize the success of the whole program. Information will soon be forgotten, and also enthusiasm for raising awareness will be lost amid a huge amount of information.

To clearly define a main message of the educational program is an excellent method to prevent this, and to focus on the essentials. This message needs to be clear, understandable, personal, and up-to-date. It should be a complete sentence, and put the whole program in a nutshell. This is the moral of the story – the main idea. It should be adopted and also transported by the guests.

When guests know the message of an educational program in advance, it is relatively easy for them to connect other information to it. This makes it easier for them to store that information in their mind, and keep everything sorted out (Ham, 1992). The whole course of the educational program is thus much easier to follow.

A clear message is also advantageous for the guide preparing a program. With the message in mind, he or she will know what kind of information is needed to get the message across to the guests. All the other information is not necessary. In this way, starting with a clearly defined message simplifies not only the planning and design of a program, but also the research and information-seeking that has to be done (Ham, 1992).

For example, from the main topic "urban tree roots", it is not clear, either to the guide or to the guests, what the program is dealing with in detail. Is it about the habit of root systems of different urban tree species, the reduced space for developing a root system in cities, or problems with soil compaction, de-icing salt, and so on? If the guests are confused about the main message, it is difficult to sort and keep information on root depth, root respiration or the uptake of nutrients. The guide preparing the program is also likely to be confused.

For an education program with an alluring title like "Back to the roots" and the clear message "Roots need to breathe. I prevent soil compaction to help trees.", this is different. Information on different substrates, their properties that will be compacted by cars, deposits or even pedestrians, and especially methods to prevent or restore this compaction, may easily be sorted. Information on different types of root systems is not necessary in this context.

20.3.4 Relevant

Education programs dealing with urban trees automatically comply with the demand to be personally relevant to the guests because they focus on something they observe every day in their immediate vicinity. Urban trees as educational objects are visible everywhere. Everybody has had his own experiences with them, positive or negative, and will meet them at every turn. To be personally and perpetually affected by an object considerably contributes to be open minded for further aspects of this topic.

20.3.5 Active

It is possible to prepare an excellent presentation with marvelous pictures to pass all the facts and information on urban trees to the audience in a classroom or lecture hall. The information is reduced in this way to solely visual, and a few acoustic, stimuli. It might even be completely reduced to visual stimuli alone, as proven by this book.

This reduction of information to just a few stimuli, or even just one, makes it more difficult for guests to keep the information in their minds for a long time. This works much better if several senses are activated though. Therefore, it is a crucial step to take the guests out to the trees, where they will encounter a range of stimuli. They see the trees, their characteristics and their environment. They experience different smells. These might be pleasant, like the smell of fresh leaves, flavorful buds or resin. Some might also be features of the environment that are awkward or embarrassing for the guests, but which are absolutely relevant for the trees, such as feces, dog urine or exhaust gases. There are sounds, including singing birds, the wind whispering in the crowns, the talking guide or noisy traffic. Finally, the guests may touch the bark or other parts of the tree, to explore their surface – smooth or rough – or even to feel different temperatures on their skin.

If all these sensations are not enough to become aware of the multiple effects of urban trees in the fields of noise and dust reduction, or temperature regulation, it is also possible to carry out data collections to quantify them (Figure 20.3). Guests can easily handle instruments to measure temperature, air humidity or noise in different situations, and draw their own conclusions on the effects of urban green (see Chapter 13). This method of education, by gathering their own data or conducting simple experiments, makes it more important and more challenging, especially for groups of adolescents or adults.

The importance of personal activity for the guests is obvious where these examples are concerned. Again, a couple of sub-competencies of the education for sustainable development are supported by gathering and possessing data and discussing the results. It is necessary to use new teaching techniques that create scenarios and situations where learners actively interact, and try to elaborate solutions for defined problems. Even more information on these teaching techniques is provided by UNESCO (2005, 2010a, 2010b, 2012).

Figure 20.3 Environmental education: together, active.

20.3.6 Entertaining

Most of the guests of environmental education programs enjoy this welcome change from everyday life. Pupils leave their classroom and school grounds. Adults or senior citizens leave their office or living room, to meet under changed conditions and work on new topics. This creates a certain familiarity and, at the same time, attention that should be used for the education program. It is also expected to have some fun, so this must be part of the program, as it helps to keep everything in mind for a long time.

The program needs to be entertaining, and there are so many easy ways to implement this: exaggerate size and time scale; use personifications; focus on an individual; tell stories, fairytales or anecdotes; and many more (Ham, 1992). Use fantasy and the imagination of the guests. For example, a role-playing game, where players assume the roles of characters in a fictional setting, is very useful in dealing with conflicting interests and demands. To represent these interests in public discussions may be a really entertaining didactic tool. It demands and stimulates various sub-competencies according to education for sustainable development. However, a little care is necessary not to make a fool of someone.

20.4 The search for professional partners

Reading this chapter, the practitioner in urban forestry, or even the decision-maker in city administrations, might be a little worried about how to organize this ambitious educational work. Hopefully, though, the value of environmental education for preservation and management of urban trees, and even the involvement of the public (Figure 20.4) becomes clear.

Figure 20.4 Joint public tree planting.

This might help to decide to support public educational institutions and, in doing so, apply the peculiarities and problems of urban trees to their work. Schools, especially in congested urban areas, are usually really interested in such cooperation. They seek for places and situations in their natural history education to experience nature personally. Only this leads to an enhanced willingness for nature conservation. The only possibility to experience nature in urban areas is in urban green (Konijnendijk, 2008).

Besides public educational institutes in every city, there are several potential partners for a joint educational work: environmental centers, NGOs, religious groups, youth clubs, specialist groups, foundations, associations, nurseries, botanic gardens, zoos, and many more. For special programs and topics, cooperation with producing or trade companies might be useful. There are almost no limits in searching for partners, but it is necessary to ensure that one's own mission is not lost in the diversity of topics and possibilities. To gain broad acceptance for urban trees and forests, their planting, development, maintenance and management, as well as the ability to recognize and balance their various functions and benefits, should be the main objective of this education.

Of course, besides cooperation with qualified partners, education and training of one's own employees and the allocation of appropriate resources is important. It even makes selection of, and communication with, partners in education much easier. As described previously, untrained persons will struggle to be able to do this ambitious work.

20.5 Conclusions

Environmental education may help to gain acceptance for the conservation and management of urban trees (Dilley and Wolf, 2013). Education for sustainable development is a perfect concept to deal with the multi-dimensional effects of trees in an urban environment.

It provides a way to assess and balance competing requirements and impacts, and develop an appropriate attitude and awareness for their sustainable management. Environmental education programs require professional and well-educated staff and partners, as well as appropriate resources.

References

de Haan, G. (2010). The Development of ESD-Related Competencies in Supportive Institutional Frameworks. *International Review of Education* **56**(2–3), 315–328 http://www.springerlink.com/content/ek411m104jwq7728/fulltext.pdf [accessed March 2015]

Dilley, J. and Wolf, K.L. (2013). Homeowner interactions with residential trees in urban areas. *Arboriculture & Urban Forestry* **39**, 267–277.

Ham, S. (1992). *Environmental Interpretation. A practical guide for people with big ideas and small budgets.* Fulcrum Publ. Golden, CO.

Jack-Scott, E., Piana, M., Troxel, B., Murphy-Dunning, C. and Ashton, M.S. (2013). Stewardship Success: How community group dynamics affect urban street tree survival and growth. *Arboriculture & Urban Forestry* **39**, 189–196.

Kohler, B. and Lude, A. (2010). *Nachhaltigkeit erleben. Praxisentwürfe für die Bildungsarbeit in Wald und Schule.* Oekom, München.

Konijnendijk, C.C. (2008). *The forest and the city. The cultural landscape of urban woodland.* Springer, New York.

Mincey, S.K. and Vogt, J.M. (2014). Watering Strategy, Collective Action, and Neighborhood-Planted Trees: A case study of Indianapolis, Indiana, U.S. *Arboriculture & Urban Forestry* **40**, 84–95.

Stoltenberg, U. (2009). *Mensch und Wald.* Oekom, München.

UNESCO (2005). *Contributing to a more sustainable future: Quality Education, Life Skills and Education for Sustainable Development.* http://unesdoc.unesco.org/images/0014/001410/141019e.pdf [accessed March 2015]

UNESCO (2010a). ESD Lens: *A policy and practice review tool. Learning and Training Tools, No.2.* http://unesdoc.unesco.org/images/0019/001908/190898e.pdf [accessed March 2015]

UNESCO (2010b). *Teaching and Learning for a sustainable Future. A multimedia teacher education program.* http://www.unesco.org/education/tlsf/ [accessed March 2015]

UNESCO (2012). *Education for Sustainable Development Sourcebook. Learning & Training Tools, No. 4.* http://unesdoc.unesco.org/images/0021/002163/216383e.pdf [accessed March 2015]

Index

Urban Tree Management: For the Sustainable Development of Green Cities, First Edition. Edited by Andreas Roloff.
© 2016 John Wiley & Sons, Ltd. Published 2016 by John Wiley & Sons, Ltd.